D1715225

COEVOLUTION

COEVOLUTION

Edited by
Matthew H. Nitecki

The University of Chicago Press
Chicago and London

MATTHEW H. NITECKI is curator of fossil invertebrates in the Department of Geology at the Field Museum of Natural History and is a member of the Committee on Evolutionary Biology at the University of Chicago.

The University of Chicago Press, Chicago 60637
The University of Chicago Press, Ltd., London

Printed in the United States of America

90 89 88 87 86 85 84 83 1 2 3 4 5

ISBN: 0-226-58686-3 (cloth)
0-226-58687-1 (paper)

LCN: 83-47773

Proceedings of the Fifth Annual Spring Systematics Symposium:
Coevolution
Held at the Field Museum of Natural History, Chicago, Illinois, on May 8, 1982
Sponsored by the Field Museum of Natural History and the National Science Foundation

CONTENTS

CONTRIBUTORS

Numbers in parentheses indicate the pages on which authors' contributions begin.

Peter Glynn (111), *Smithsonian Tropical Research Institute, APO, Miami, Florida, 34002*

Simon A. Levin (21), *Division of Biological Sciences, Section of Ecology & Systematics, Cornell University, New Biology Bldg., Tower Road, Ithaca, New York, 14850*

Everett C. Olson (307), *Department of Biology, University of California, Los Angeles, California, 90024*

Douglas W. Schemske (67), *Department of Biology, Barnes Laboratory, University of Chicago, 5630 South Ingleside Avenue, Chicago, Illinois, 60637*

Montgomery W. Slatkin (339), *Department of Zoology, NJ-15, University of Washington, Seattle, Washington, 98195*

Robert M. Timm (225), *Department of Zoology, Field Museum of Natural History, Roosevelt Road at Lake Shore Drive, Chicago, Illinois, 60605*

Diana F. Tomback (179), *Division of Natural and Physical Sciences, University of Colorado at Denver, 1100 Fourteen Street, Denver, Colorado, 80202*

Leigh M. Van Valen (1), *Department of Biology (Whitman), University of Chicago, 915 East 57th Street, Chicago, Illinois, 60637*

S. David Webb (267), *Florida State Museum, Museum Road, Gainesville, Florida, 32611*

PREFACE

Plant and animal biologists have a common interest in the evolutionary origin of species. They therefore share problems and methodologies, including the need to interpret and use information from such fields as population genetics, cytology, biochemistry, ecology, molecular biology, animal behavior, and paleontology. A dialogue among all biologists furthers science by permitting an exchange of data, methodologies and ideas among workers facing a commonality of questions and procedures.

At present there are very few conferences directed to biologists working on diverse taxonomic groups. Those that do exist are either very local in scope or else they tend to be dominated by either botanists or zoologists. Rarely is an attempt made to include paleontologists or anthropologists as well. The Field Museum Spring Systematics Symposium fulfills a need not yet met by any other meeting, namely the provision for a regular dialogue among biologists that contributes to the work of all the diverse disciplines concerned with evolution.

Within the last two decades, coevolution has been widely recognized as an important area of research. The present volume treats the topic of coevolution at the level of (1) whole communities, particularly in terms of predator-prey relationships, general competition and character displacement; (2) small groups of species with emphasis on host-parasite relationships, allelopathy, chemical defenses and plant toxins; and (3) two interacting species. Although there have been a number of conferences on coevolution, our symposium is unique in that it features recent theoretical models that treat genetic aspects of coevolution and considers coevolution within neontological, as well as paleontological aspects. The organization of the book is as follows: two introductory papers (Van Valen and Levin) are succeeded by discussions on coevolution in plant-animal communities (Schemske); corals and corallivores (Glynn); birds and pines (Tomback); lice and gophers (Timm); the evidence of coevolution from the fossil record (Webb and Olson); and the applicability of the coevolutionary model (Slatkin). Agreement on the definition of coevolution has not been reached. Thus Van Valen discusses how coevolution may in effect mean much of evolution. Notably, he does not restrain coevolution narrowly to the inter-specific associations. Levin defines coevolution within the framework of genetically explicit models between pairs of closely associated species and the dynamics of coevolution within interacting populations. Schemske, in his genetical and ecological approach, radically separates coevolution from mutualism and interactions. Glynn, within the complicated nature of his data, defines coevolution in ecological terms almost as mutualism. Tomback also deals with mutualistic associations, and her definition of coevolution requires evaluation of each mutualistic system, including coadaptation. Timm builds unique, strong and presumptive evidence for coevolution of host and parasite. Webb argues for the prominence of coevolutionary relationships in living and fossil mammalian assemblages. Olson, with insight, asks whether the genera of Permian reptiles underwent coevolution or coadaptation. Finally Slatkin applies mathematical models to the concept of ecological character displacement, the most fundamental ecological component of community evolution.

Thanks are due to those who reviewed drafts of the individual chapters: Donald P. Abbott, Andrew J. Beattie, William C. Burger, Robert L. Carroll, Paul K. Dayton, John W. Fitzpatrick, Philip D. Gingerich, Lynne D. Houck, Ke Chung Kim, George V. Lauder, Yan B. Linhart, Kubet Luchterhand, Robert M. May, Michael Nee, Bruce Patterson, Roger D. Price, Leonard Radinsky, Daniel Simberloff, Alan Solem, Roger D.K. Thomas, John N. Thompson, William D. Turnbull, Peter P. Vaughn, Michael J. Wade, and Rupert L. Wenzel. Zbigniew Jastrzebski counseled on and helped with many text illustrations, and Rosanne Miezio prepared figures 6 and 7 of Webb's chapter. Laurel Johnson proofread many a paper and the index. Clarita M. Nunez, with celestial patience, art, and speed marshalled all the undecipherable intricacies of camera-ready typing, editing and correcting.

Finally the National Science Foundation (DEB 8023139) and the Field Museum lent financial wings to us. To all these, I pay my editor's debt.

HOW PERVASIVE IS COEVOLUTION?

Leigh M. Van Valen

Department of Biology
University of Chicago
Chicago, Illinois

Coevolution occurs on a very broad scale and may comprise most of evolution. At the community level the limitation of available free energy promotes coevolution. Coevolution commonly occurs as a result of interaction among different levels, time scales, and areas of natural selection. Not all levels commonly distinguished are selectively distinct, and couplings among levels are often confused with phenomena within levels. Patch selection may cause previously puzzling phenomena, but neither it nor any other cause is available for the Gaia hypothesis. Despite its predictive successes, Gaia fails as a scientific explanation.

Coevolution occurs when the direct or indirect interaction of two or more evolving units produces an evolutionary response in each. Even so restricted, coevolution may comprise most of evolution. This possibility is not apparent in the common aphorism that evolution occurs by adaptation to the environment, for we then tend to think of "the environment" as mostly physical. Biotic influences, even interactions, can often produce a response in only one of the species involved, a one-way effect like that of physical influences. However, I will try to indicate that the multifarious aspects of coevolution spread much more widely than is usually believed.

Contrarily, not everything that looks like coevolution is really coevolved, although the distinction may often be difficult to make in practice. As H. G. Baker (in conversation) and Janzen (1980) have emphasized in part, coevolution can be mimicked by such things as sequential adaptations from different causes or simultaneous adaptations to the same environment.

Discussion of coevolution has for the most part been restricted to some rather specific cases, such as pollination adaptations, adjustments to competition, specific mutualisms, predator-prey arms races, and herbivory. Mutualistic evolution benefits both (or all) participating species in relation to the rest of the community, while the other processes do not. Take for instance a common pattern of evolution in fungal or nematode or insect parasitism of plants (Day, 1974; Ellingboe, 1979; Sidhu and Webster, 1981). A grass may evolve resistance to a rust by a single-gene substitution. The rust, in turn, counters this gene by single-gene substitution of its own. The grass counters this with another, and so on, perhaps eventually even forming a cycle if reversals occur. Other patterns also occur (Nelson, 1979; Segal et al., 1980), such as polygenic resistance and a stable low-level susceptibility. (Is the difference here meaningful, more than the historical accident responsible for similar differences in insecticide resistance?) However, the evolution of resistance has a cost, as indicated by some ecological regularities (Rice and Westoby, 1982).

Many parasites are sufficiently host-specific that they can help in deciphering the phylogeny of their hosts, if appropriate caution is exercised. This is true for both plants (Hedberg, 1979) and animals (Brooks, 1979). When each adapts to the other, as is usual, coevolution occurs at a larger scale.

COEVOLUTION AND COMMUNITIES

The common currency of evolution is free energy (Van Valen, 1976a). All aspects of evolution fall easily into this context, and some critical cases seem to be interpretable in a coherent way only in this context. Moreover, energy is causal, unlike such things as biomass or numbers of "individuals" or efficiency.

In almost all cases, on both ecological and evolutionary time scales, the free energy potentially available to an organism is both limited and potentially available to other organisms. If so, competition is nearly ubiquitous, although it can be indirect (as by mutualisms) or diffuse (as for most plants) and need not operate in realized form either continuously or strongly. In this view a predator competes with its prey for use of the free energy in the prey.

For plants, the availability of free energy is commonly limited by the imperfections of physiology. In a mature forest it may almost all be used, but in a desert little is. A hypothetically perfect physiology would make photosynthetic factories of deserts, and indeed natural selection operates on desert plants to bring them as close as they can come to this perfection. But it isn't very close. This is, of course, because water is necessary and scarce. Other things are scarce elsewhere, especially for the third of the world's photosynthesis done by aquatic and algae plants, but on land the availability of water seems to be much the most important control of productivity (Rosenzweig, 1968).

For animals and decomposers, including chemoautotrophic bacteria except at some hydrothermal rifts, the energy plants and algae fix in their production is what is potentially available. The heterotrophs succeed in obtaining more than 99 percent of it (Van Valen, 1976a), and almost all the rest is buried in sediments; the successful reproduction of plants (e.g., the average replacement of one tree by one successful seed before the seed's own photosynthetic production, plus the respiration of the eventually unsuccessful seeds and pollen) is energetically negligible.

Plant-sucking homopterans and a few other animals have an excess of energy which would be available to them if they could use it, as for desert plants. Like them also, an imperfect physiology prevents full use of the energy but they are always selected to make more of it available if they can. This is true however the density of an animal population (or metapopulation) happens to be regulated; if there are more animals they use more energy.

Thus we find, to a close approximation, a zero-sum constraint on the absolute fitnesses in a community. This is a strong constraint which I have called the Red Queen's Hypothesis (Van Valen, 1973) and implies much coevolution as well as competition. If species A increases its fitness, then one or more other species B lose fitness, by the same total amount. This effect in all likelihood will change the distribution of selective vectors on species B. By doing so it upsets their quasi-equilibrium (multivariate stabilizing selection) and can therefore be expected to permit directional response to selection. Such a response by species B will usually increase the fitness of species B, because cases where good of the "individual" differs from good of the local population of its species are rare. Then the whole process starts again, for the increase in fitness of B reverberates onto other species. If B does not increase in fitness it has taken a step toward extinction, which itself is a kind of evolution.

Sometimes, when a species is introduced into a new area there is no detectable change in abundance or phenotype of more or less similar species. What does this tell us? I think it tells very little at the community level. The new species is eating something; what ate this food before? Maybe nothing much did until it died and became food for decomposers. But even in such a case there is now less food for these decomposers; the new species will degrade part of the free energy in this food and perhaps make the rest it eats unavailable to the previous decomposers. So we must beware of tests that don't test; I have discussed elsewhere the matter of testing deductively based conclusions (Van Valen, 1976b, 1982).

The very possibility of large-scale community patterns is shocking to some ecologists. But if we look for them we may find them. One that is now emerging concerns the relation between body size and energy flow. For both mammals and birds, with poorer data suggesting greater generality, density decreases with body size in a way that makes the energy flow per species per area constant over all body sizes. There is of course much scatter, but the overall trend is clear. The mammal result is from Damuth (1981) and the others are from independent work of mine.

Why should species have their densities regulated in a way that produces this pattern? The answer isn't known but presumably involves coevolutionary interactions as defined at the beginning of this paper. I suspect that the mechanism differs between high and low deviations. A species much more energetically dense than most is more vulnerable in two ways: as an unusually valuable potential source of food for parasites and predators (Maiorana, 1979) and in many or most cases as occupying, in part perhaps by a spillover effect (Van Valen, 1976a), unusually broad niches. Thus its very density increases the probability of detrimental coevolutionary responses by other species.

For unusually rare species the reverse of the above effect should operate somewhat, but perhaps more importantly an energetically rare species can more readily become extinct, at least for a given community. This is because the same absolute encroachment has a greater effect on a rare species than on a common one. That this effect can be important is suggested by the fact that hummingbirds, adaptively isolated from the central granivorous-insectivorous cluster of terrestrial birds, can get very rare energetically (low energy control by the species per unit area) but nevertheless persist.

It is advantageous for an evolutionary unit to change the mean or variance of its body size only if more expansive energy (or energy used for growth and reproduction) can be obtained at some other distribution of body size. In cases where body size is important in resource partitioning such a change may usually have to await a decrease in the fitness of some other species, perhaps resulting in the local extinction of that species. Such parameters as the stability and predictability of energy can strongly affect its availability to a given species, and populations may be established in environments of lower available energy, but an evolutionary response occurs almost only (Van Valen, 1976a) when more expansive energy is obtained as a result of the evolution.

LEVELS OF COEVOLUTION

It should be axiomatic that the response to selection at one level or time-scale increases fitness at that level or time-scale beyond what it would be without selection, and that any increase of fitness at other levels or time-scales is an accidental by-product not to be expected unless the levels are causally coupled in a suitable way. (Like most other axioms this has somewhat less than a universal domain, but the only exception I know is for environments that cycle out of phase with selection.) A level is the set of units, such as alleles, individuals, or lineages, among which a particular kind of natural selection operates.

Often there is such a causal coupling. Most selection at the individual level increases the fitness of the population, as formalized in Wright's principle of the maximization of $\overline{W}$ (his population fitness) or as seen from the natural-history viewpoint in that the great majority of adaptations which benefit an individual also benefit its population. A similar coupling occurs on different time-scales of selection; what is advantageous in one generation will usually be advantageous in ten, a thousand, or a million, although of course the strength of the coupling decreases with time as it does with space.

The domains of couplings are narrower than those of maximization within single levels. For instance, frequency-dependent selection need not maximize $\overline{W}$ (Wright, 1969) although what may be the most common frequency-dependent cases, heterosis and suitably caused stabilizing selection, do so.

It is worth noting that evolutionary stable strategies, sometimes advocated as a replacement for any maximization when the latter concept is restricted to couplings, are a subset of single-level maximizations. Moreover, the term is somewhat of a misnomer, because (like any other phenotypic character) their evolutionary stability can vanish with a change in the evolutionary context (Van Valen, 1980a).

Other interactions among levels differ from the reciprocally reinforcing ones I have called couplings. Consider the relations between cells and the individuals they comprise.

Here the adaptive relation is one-way: cells exist for the sake of their organisms, but not *vice versa*. The reproduction of cells is not autonomous, even for a sponge or a filamentous alga, but is controlled at a higher level. Cells of one kind, or in one region, can be caused to stop dividing or even to die. Eventually the whole soma always dies unless there is asexual reproduction.

And the cells of an individual often, perhaps almost always, differ in their genotype. Apart from somatic mutation and aging effects, differentiation itself proceeds in part by genotypic changes such as differential replication of DNA, DNA methylation, and even, more rarely, gross chromosomal changes. The activation or repression of genes in a cell is commonly inherited by the descendants of the cell. Conversely, corresponding cells in different individuals may have more in common than divergent cells in the same individual. This is undoubtedly the case where there has been asexual reproduction, as with identical twins or grass ramets.

There can nevertheless be coevolution between cells and individuals, just as among cells themselves. If a cell successfully escapes the controls imposed on its own reproduction it forms a tumor. More of the individual than before is now composed of this kind of cell; it has been selected for. In animals, where whole cells can move, some may invade adjacent tissue; such a tumor is called a cancer. Animals have various kinds of defenses against cancers and other tumors (cf. Cairns, 1975); such defenses close the causal loop and thereby complete a coevolutionary circuit. Unlike most coevolutionary circuits, this ends in one loop because of the discontinuity of different instances of cancer and their resulting inability to respond to selection by their host's defenses, except for metastases, caused by selection among the cancer cells themselves, in single hosts.

When there is selection in opposite directions at different levels or time-scales a coevolutionary equilibrium among alleles or other units can occur in a substantial proportion of realistic cases. However, we have to be careful that the levels or time-scales are different enough. For instance, meiotic drive and gametic selection are phenomenologically quite distinct processes and either can produce an equilibrium with individual selection. They cannot, however, produce an equilibrium with each other in one sex; like the mortality, fertility, and time components of individual selection, their effects merge into a single resultant vector. (The vector may select for an intermediate phenotype, but this result has no relation to the equilibrium of alleles produced by discrete vectors.)

Similarly, sexual selection seems different enough from other kinds of individual selection that Darwin distinguished them sharply. If we look at sexual selection causally, though, it is merely one of the components of the fertility part of individual selection. A peacock's tail is a result of the addition (on the proper scale) of the component selective vectors, just as are the number and size of flowers on a Queen Anne's lace or the size of the human birth canal. There is no way a population can become extinct as a byproduct of any kind of sexual selection itself (unless nearly all the sexual winners of one sex are sterile), but sexual selection can lower average viability and so make a population more susceptible to extinction. Other components of the fertility and viability vectors can do so too; fertile female *Drosophila* die sooner than sterile ones or virgins, and the common specialization of a parasite from several host species to a single one increases its probability of extinction. That sexual selection can also lower the competitive ability or energy control of a population is just another sort of restriction on the domain of coupling of individual and population selection.

If selfish DNA exists, and it probably does (Doolittle, 1982), some of it as parasitic chromosomes (White, 1973), selection within the genome can produce coevolutionary equilibria with any other level. The possibility of such equilibria has been discussed only with respect to individual selection. Gametic selection is at least equally plausible, in the absence of evidence, to balance the nucleotide or chromosome selection of selfish DNA. Conceivably the increase of selfish DNA could even contribute to lineage extinctions by interfering with response to individual selection.

Local competition among species is a form of natural selection, even if we usually consider it separately, and this too can lead to coevolutionary equilibria as well as directional elimination. Besides local niche adjustments, regulation of density and thus the potential for competition sometimes occurs at a regional level, for species adapted to ephemeral habitats or early succession, or for fugitive species. They affect not only each other but often more locally stable species, thereby closing the coevolutionary loop.

By a little extension we can consider coevolutionary spirals. Here species A affects species B but then becomes locally extinct, perhaps from the invasion of a similar species C. C then is affected by B's response to A even though it may itself produce no further response. Here the coevolution is of B with a set of mutually similar species, but there may be no closed loop among the same species.

Natural selection operates on all time scales together as well as on all levels, and any two time scales, like any two levels, can produce a coevolutionary equilibrium. In fact time scales are not sharply distinct from the higher levels (Van Valen, 1976a). The only documented examples I know are important aspects of the regulation of the distribution of body size of the Mammalia and the Foraminifera. In both groups geologically short-term selection tends to increase body size and long-term selection decreases it, although by different mechanisms for the two groups (Van Valen, 1975). Reversal of the direction of selection at different time scales (*not* the same as selection that fluctuates in time) is undoubtedly rather common, although it is difficult to find characters with enough data for analysis. Of course short-term selection can be strong enough that no variation remains for long-term selection to act on, in which case no equilibrium is possible.

PATCH SELECTION

There is a third sort of unit of selection, besides those levels and time-scales, and this is spatial. Although it has hardly begun to be studied its requirements for operation are sufficiently broad that it may eventually prove to be among the most prevalent coevolutionary mechanisms. Wilson (1980), who discovered it, calls it trait-group selection; I prefer the mechanistically more descriptive term patch selection (Van Valen, 1980b).

A patch is a region where individuals affect each other, with its contained individuals. Patches may be overlapping parts of a continuously inhabited area. Commonly there is variation, random or otherwise, among patches. Such variation can cause differential productivity of the patches. After this selection among patches by differential productivity, phenotypes and genotypes which happen (even randomly) to be more prevalent in the more productive patches than elsewhere will automatically have increased in relative frequency overall.

Thus patch selection works even with complete mixing among patches every generation.

Note that there is no requirement that the members of a patch be of the same species, although they may be or one's interest may be restricted to one species. To take a recent problem, Price at al. (1980) note that it would be advantageous to plants to help the parasitoids of their own herbivores but realize, with Williams (1966) in a different context, that individual selection cannot do this if parasitoids move much among genets (sexually produced individuals or clones) of plants. They conclude that such evolution is unlikely to occur. However, patch selection makes evolution like this easy. The existence of a broadly applicable mechanism should motivate search for real examples of this and other phenomena which have been theoretically forbidden except as accidents.

For instance, patch selection may well have been involved in an unusually realistic laboratory experiment by Pimentel, Nagel, and Madden (1963). Flies and parasitoid wasps were allowed to evolve in a weakly interconnected matrix of population cages. Not surprisingly, the flies increased in average resistance. However, the wasps evolved toward lower reproduction per individual. This extraordinary result may come from the advantage a low-fertility female would have when immigrating to a new cage with flies but no other wasps. By allowing enough flies to survive to produce a large population for her daughters, she would have many grandchildren. This may not have been what really occurred, but the experiment has been one of the minor mysteries of genetics. This work should be repeated and extended.

Beyond the patch, which may be as small as two individuals, the only other spatial unit of selection is probably that of the province, continent, or other more or less isolated piece of land or body of water. For instance, island animals are commonly said to be inferior to mainland ones, and indeed most known extinctions in the past few centuries have been on islands. However, the appropriate comparison would involve relative local survival of introduced and indigenous species in a similar frequency of introductions from islands to mainland, and this experiment has not occurred. Because islands are small they have relatively few species, and we need to know how the probability of a random island species doing well on the mainland compares with the reverse. For instance, a major radiation of the Drosophilidae probably started on the Hawaiian Islands and spread to the mainland (Throckmorton, 1967). Even the invasion of South America by North American mammals is much less one-sided than is commonly thought. The probability for a family or a genus to emigrate was the same in each direction (Marshall, 1981; Marshall, Webb, Sepkoski, and Raup, 1982); in fact more families migrated north than south. However, the megafaunal extinction 10,000 years ago somehow affected taxa of South American origin much more severely than it did those of North American origin, and more South American taxa (including large predaceous birds) had earlier become extinct because of the competitive effects of a greater radiation of North American emigrants. Thus although large-scale coevolution undoubtedly occurs it is unclear whether its rate frequently differs among different geographic regions.

GAIA

The grandest form of coevolution that has been proposed is the Gaia hypothesis (Lovelock, 1972, 1979; Margulis and Lovelock, 1974). It is a seductive hypothesis but I believe it is wrong. This view needs a somewhat detailed presentation because of the importance of the hypothesis and because all comments on Gaia I have seen have been poorly focused.

There are many geochemical and even geophysical phenomena of global scale which are, or may be, regulated by processes of living organisms. By regulation I mean feedback control, so that departure in either direction from the control point will induce or increase an opposing process. The control points happen to be suitable for life as we know it, and the major departures from them which would occur without organismal regulation would be lethal for most or even all organisms. The departures would in some cases be several orders of magnitude.

The best known of these phenomena is the amount of molecular oxygen in the atmosphere, which is now just below the level where even wet vegetation sustains catastrophic fires. Other phenomena include the temperature at the surface of the Earth, removal of toxic metals from the ocean, restoration of elements from the sea to the land, the making available of useful elements, a neutral pH in the oceans and in rain, the concentration of oceanic salts, ozone concentration and the resulting shield from ultraviolet, and the amount of molecular nitrogen in the atmosphere with the resulting atmospheric pressure and dilution of O_2.

The Gaia hypothesis is more than a collection of such feedback systems. Its central claim is that the feedbacks are adaptive, controlled "by Darwinian natural selection" in some way which remains to be clarified. The adaptation is for "the maintenance of conditions optimal for life in all circumstances" (Lovelock, 1979, p. 128).

Although there is not general agreement in all cases on the controlling mechanisms of the dynamic equilibria or the participation of organisms in them (Holland, 1978; Walker, 1977; Schopf, 1980) this is not the basis of my critique. Neither is the likelihood that life on Earth has adapted to the changes life itself has caused, such as the first great pollution of the Earth (in Maiorana's phrase) by O_2. I claim instead that the proposed general mechanism does not work.

It is dangerous to deny a phenomenon by claiming that there can be no mechanism for it. Lyell and Darwin, not Kelvin, were right with respect to the age of the Earth, Wegener and Holmes were right with respect to continental drift, and so on. Nevertheless this is the basis of my denial. The danger is particularly acute when, as in the cases I mentioned, there is much supporting evidence for the phenomenon and the evidence is not otherwise accounted for. This is not the case for Gaia because there is a competing hypothesis of comparable generality.

The competing hypothesis is that, in the absence of a suitable environment, we would not be here to think about it. It is called the anthropic principle in physics. In this view Gaia exemplifies what may be called Henderson's fallacy, like the bias of ascertainment in human genetics. Lawrence Henderson (1913) wrote the first extensive treatment of the Gaia approach, although it was much less explicit than Lovelock's and took the viewpoint that all or most aspects of the environment, including the properties of the chemical elements, were adapted to life just as life was adapted to them. (This view, in turn, was derived from the pre-Darwinian natural theology.)

Henderson's fallacy is that conditions necessary for a phenomenon provide evidence that they were themselves produced or modified by that phenomenon. So stated it sounds faintly ridiculous, but that is, I think, a fair if bald restatement of the argument underlying Gaia. I do not want to deny that organisms control much of the geochemistry of the Earth's surface, although how much they control is controversial. This subject is an important one and its investigation needs encouragement rather than attenuation. However, the Gaia hypothesis is not that organisms have large geochemical effects. It is rather that these effects have evolved as adaptations themselves, not as side effects of more local adaptations. This causal hypothesis of otherwise unknown natural selection is what I am discussing.

I said that there is no mechanism for Gaia. Natural selection must be selection *among* something, yet Gaia is unitary by hypothesis. How did it (or she) evolve, if not by intelligent self-modification? We might think that species contributing to Gaia would be favored over detrimental species, but there is no functional connection between contributing to Gaia and benefiting from her. Rather the reverse. If methanogenic bacteria somehow manage to increase their output when the amount of atmospheric O_2 increases (the reverse, incidentally, of the direction of change a causal analysis suggests), then they would not be among the beneficiaries when it declines again. Such inverse effects would be rather prevalent if Gaia exists.

Alternatively we can consider patch selection. This looks more promising, because it can lead to adaptive modification of geochemical processes (Wilson, 1980). The problem here is of the scale of feedback. For most proposed aspects of Gaia the feedback loop is nearly or quite global, so a patch which is a benefactor will get no special advantage from being a benefactor. It thus will not be selected for.

Perhaps then long-term selection might work: if a biota didn't contribute to Gaia it would eventually disappear. Stated in this way it merges with the anthropic principle rather than with Gaia; only suitable biotas survive. To generate or maintain Gaia selectively there must be co-existing or sequential biotas with different degrees of contribution to Gaia. Perhaps the great Permian extinctions were caused mostly by the sea becoming brackish as a result of the great volumes of evaporates deposited then, probably even more evaporites than we know if many have since disappeared. This would be a failure of Gaia by the Paleozoic marine biota. What did such a failure select for? There was no suitable variation available for it to act selectively on, and the Mesozoic must just try again. If so, the extinction itself was biotically selective, but if there are many aspects of Gaia the selection on just one aspect was so weak as to be useless overall, even in geological time.

Considered just as a hypothesis, Gaia makes a variety of confirmed predictions of considerable scope. If we judge hypotheses by their predictive power and success, as some narrow-minded people would have us do, we should embrace our mother Gaia. Considered causally, though, Gaia fails, and I think this failure is decisive.

Thus coevolution is pervasive, but it is not as pervasive as we can imagine.

ACKNOWLEDGEMENTS

I thank R.D.K. Thomas and another referee for their comments.

LITERATURE CITED

BROOKS, D.R. 1979. Testing the context and extent of host-parasite coevolution. Systematic Zoology, 28:299-307.

CAIRNS, J. 1975. Mutation selection and the natural history of cancer. Nature, 255:197-200.

DAMUTH, J. 1981. Population density and body size in mammals.

Nature, 290:699-700.
DAY, P.R. 1974. Genetics of Host-Parasite Interaction. San Francisco: Freeman. 238 pp.
DOOLITTLE, W.F. 1982. Selfish DNA after fourteen months. *In* Dover, G.A. and R.B. Flavell (eds.), Genome Evolution, pp. 3-28. New York: Academic Press.
ELLINGBOE, A.H. 1979. Inheritance of specificity: the gene-for-gene hypothesis. *In* Daly, D.M. and I. Uritani (eds.), Recognition and Specificity in Plant Host-Parasite Interactions, pp. 3-17. Baltimore: University Park Press.
HEDBERG, I. ed. 1979. Parasites as plant taxonomists. Symbolae Botanicae Upsalienses 22(4):1-221.
HENDERSON, L.J. 1913. The Fitness of the Environment. New York: Macmillan. 317 pp.
HOLLAND, H.D. 1978. The Chemistry of the Atmosphere and Oceans. New York: Wiley. 351 pp.
JANZEN, D.H. 1980. When is it coevolution? Evolution, 34:611-612.
LOVELOCK, J.E. 1972. Gaia as seen through the atmosphere. Atmospheric Environment, 6:579-580.
LOVELOCK, J.E. 1979. Gaia: A New Look at Life on Earth. Oxford: Oxford University Press. 157 pp.
MAIORANA, V.C. 1979. Nontoxic toxins: the energetics of coevolution. Biological Journal of the Linnean Society (of London), 11:387-396.
MARGULIS, L., and J.E. LOVELOCK. 1974. Biological modulation of the Earth's atmosphere. Icarus, 21:471-489.
MARSHALL, L.G. 1981. The great American interchange - an invasion-induced crisis for South American mammals. *In* Nitecki, M.H. (ed.), Biotic Crises in Ecological and Evolutionary Time, pp. 13-229. New York: Academic Press.
MARSHALL, L.G., S.D. WEBB, J.J. SEPKOSKI, JR., and D.M. RAUP. 1982. Mammalian evolution and the great American interchange. Science, 215:1351-1357.
NELSON, R.R. 1979. Some thoughts on the coevolution of plant-pathogenic fungi and their hosts. *In* Nickol, B.B. (ed.), Host-Parasite Interfaces, pp. 17-25. New York: Academic Press.
PIMENTEL, D., W.P. NAGEL, and J.L. MADDEN. 1963. Space-time structure of the environment and the survival of parasite-host systems. American Naturalist, 97:141-167.
PRICE, P.W., C.E. BOUTON, P. GROSS, B.A. MCPHERON, J.N. THOMPSON, and A.E. WEISS. 1980. Interactions among three trophic levels: Influence of plants on interactions between insect herbivores and natural enemies. Annual Review of Ecology and Systematics, 11:41-65.
RICE, B., and M. WESTOBY. 1982. Heteroecious rusts as agents

of interference competition. Evolutionary Theory, 6 (in press).

ROSENZWEIG, M. 1968. Net primary productivity of terrestrial communities: prediction from climatological data. American Naturalist, 102:67-74.

SCHOPF, T.J.M. 1980. Paleoceanography. Cambridge: Harvard University Press. 341 pp.

SEGAL, A., J. MANISTERSKI, G. FISCHBECK, and I. WAHL. 1980. How plant populations defend themselves in natural ecosystems. *In* Horsfall, J.G. and E.B. Cowling (eds.), Plant Diseases, vol. 5, pp. 75-102. New York: Academic Press.

SIDHU, G.S., and J.M. WEBSTER. 1981. The genetics of plant-nematode parasitic systems. Botanical Review, 47:387-419.

THROCKMORTON, L.H. 1967. The relationships of the endemic Hawaiian Drosophilidae. Studies in Genetics (University of Texas) 3(for 1966):335-396.

VAN VALEN, L.M. 1973. A new evolutionary law. Evolutionary Theory, 1:1-30.

VAN VALEN, L.M. 1975. Group selection, sex, and fossils. Evolution, 29:87-94.

VAN VALEN, L.M. 1976a. Energy and evolution. Evolutionary Theory, 1:179-229.

VAN VALEN, L.M. 1976b. Domains, deduction, the predictive method, and Darwin. Evolutionary Theory, 1:231-245.

VAN VALEN, L.M. 1980a. Evolution as a zero-sum game for energy. Evolutionary Theory, 4:289-300.

VAN VALEN, L.M. 1980b. Patch selection, benefactors, and a revitalization of ecology. Evolutionary Theory, 4:231-233.

VAN VALEN, L.M. 1982. Why misunderstand the evolutionary half of biology? *In* Saarinen, E. (ed.), Conceptual Issues in Ecology, pp. 323-343. Dordrecht: Reidel.

WALKER, J.C.G. 1977. Evolution of the Atmosphere. New York: Macmillan. 318 pp.

WHITE, M.J.D. 1973. Animal Cytology and Evolution, edition 3. Cambridge: Cambridge University Press. 961 pp.

WILLIAMS, G.C. 1966. Adaptation and Natural Selection. Princeton: Princeton University Press. 307 pp.

WILSON, D.S. 1980. The Natural Selection of Populations and Communities. Menlo Park: Benjamin/Cummings. 186 pp.

WRIGHT, S. 1969. Evolution and the Genetics of Populations. Vol. 2. Chicago: University of Chicago Press. 511 pp.

SOME APPROACHES TO THE MODELLING OF COEVOLUTIONARY INTERACTIONS

Simon A. Levin

Section of Ecology and Systematics
and
Ecosystems Research Center
Cornell University
Ithaca, New York

The role of models in the understanding of coevolutionary interactions is explored, ranging from explicit genetic models to purely phenotypic ones. Special attention is devoted to gene-for-gene plant-pest resistance systems, the evolution of avirulence in host-parasite associations, and the evolution of aspect diversity in predator-prey communities.

*Explicit genetic models are most appropriate to small ensembles of tightly interacting species in which the genetic basis of change is well understood. The gene-for-gene systems of cereal plants and their fungal pathogens are ideal in this regard. However, most classical models ignore ecological and epidemiological interactions, which are critical to the understanding of phenomena such as the evolution of reduced virulence in parasite-host associations and the stabilization of such associations at intermediate levels. Models which incorporate these elements are discussed in some detail, with special reference to extensions of the work of Levin and Pimentel (1981) on the interaction between the myxoma virus and the Australian rabbit (*Oryctolagus cuniculus*).*

Diffuse coevolution, involving many species, requires yet a different perspective. The work of Levin and Segel (1982) on the evolution of aspect diversity in predator-prey communities is discussed, and the critical issues identified. Extensions to other systems are suggested.

INTRODUCTION

Models of evolutionary and coevolutionary processes, just as models in so many branches of science, serve primarily to aid our understanding and to provide a framework within which to couch questions. As Jacob (1977) points out, the process of evolution is more the work of a tinkerer than that of a master craftsman. The influences of the historical record and stochastic events predominate. Hence, whereas explanation is an achievable goal, prediction is often impossible.

There are exceptions: the development of heavy metal tolerance and pesticide resistance are predictable responses to stresses, and even estimates of the time to develop resistance can sometimes be forecast with reasonable accuracy (B. Levin et al., 1982). But these relate to single factor influences. What makes evolutionary prediction so difficult is the coevolutionary context, including both interspecific and frequency-dependent intraspecific effects. Because of the complexity of interactions within ecosystems, the problem of predicting changes is a vexing one even on an ecological time scale. Prediction of evolutionary change is even more refractory. The difficulties are basic and inherent in the nature of large-scale nonlinear systems with complex linkages; their dynamic behavior is typically erratic and highly sensitive to slight parameter changes. Deviations in the behavior of initially similar systems may become magnified as effects are propagated through the network of system interactions and feedbacks; this makes detailed forecasting a virtually impossible chore. On the other hand, the retrospective approach has been a very profitable one in evolutionary theory, and has led to great advances in our understanding.

POPULATION GENETICS AND EVOLUTIONARY THEORY

Much of the classical theory leading to the modern synthesis of population genetics and evolutionary theory emerged from beginnings in plant and animal breeding, in which selection regimes were generally single factor and were well understood because they were imposed by the breeder. Within the context of artificial selection, remarkable strides were made in bridging the gap between a phenotypic view of evolution and one based on genotypic proportions; but in the study of evolution in natural settings, the problems remain severe (Lewontin, 1980). In fact, Mayr (1982; see also discussion in Lewin, 1982) argues that population geneticists present a reductionist view of evolutionary change which is an impediment to understanding.

Important among the contributors to the synthesis of the 1930's were three men -- Sewall Wright, Ronald Fisher, and J.B.S. Haldane -- who developed the mathematical theory and explored its ramifications. Among the major achievements of the theory were ways to relate population-level properties, such as changes in mean fitness, to events at the individual level. The search for macroscopic parameters and the description of their dynamics were and remain central problems in evolutionary theory. Fisher's Fundamental Theorem of Natural Selection and Wright's Adaptive Surface together tell the story of the population's gradual progress through inferior regimes of mean fitness until some pinnacle of fitness is achieved, albeit perhaps not the highest of pinnacles. This paradigm has undoubtedly been among the most powerful in evolutionary theory, and has contributed to the rise of optimization theory in evolutionary ecology.

But it is a misleading paradigm. The conclusion that populations evolve towards maximization of mean fitness is easily vitiated, and the worst culprit is frequency dependence (see Levin, 1978). Indeed, there are other impediments -- environmental change, linkage, epistasis, and density dependence. But provided the effects of these complications are not too severe, weakened versions of the Fundamental Theorem can be constructed. In theory that is also the case with frequency dependence; but, where ecological interactions are involved, frequency-dependent effects must eventually become severe as populations reach the limits of available resources. By definition, a population in near steady state has mean fitness near 1, and evolution cannot improve on that over the long term. An evolving population necessarily is genetically heterogeneous, and one in which the frequencies of genotypes are changing. As a consequence of this change, the environmental milieu also shifts. Thus, even in the intraspecific case, the population is coevolving with its environment. This leads to nonlinear feedbacks which confound the simplistic theory. Indeed, theoretical examples involving frequency dependence can be constructed in which mean fitness is minimized rather than maximized at equilibrium.

Theorists delight in the demonstration of such problems, for they show that the mathematical theory is not tautological. The paradox arises because, while Nature selects the most fit within a given environmental context, simultaneously that context changes (coevolves) as a consequence of selection. Gould (1977) points out, "Natural selection is a theory of *local* adaptation to changing environments. It proposes no perfecting principles, no guarantee of general improvement." Lewontin (1977) succinctly summarizes, "Adaptation, for Darwin, was a process of becoming rather than a state of final optimality."

THE CLASSICAL THEORY APPLIED TO INTERACTING POPULATIONS

Wright set forth a discrete-time framework (summarized in Wright, 1955) for describing evolutionary change, one which has influenced the mathematical development ever since. This framework extends naturally to systems of interacting populations and to changing environments, including effects resulting from frequency and density dependence. Mode (1958) pioneered models of coevolving populations in considering the gene-for-gene resistance systems of cereal grasses and their associated pathogens. Subsequently, numerous others have studied similar genetic models (e.g., Jayakar, 1970; Yu, 1972; Levin and Udovic, 1977; Levin, 1978). Levin and Udovic (1977) summarize the complexity of possible relationships when gene frequencies and population densities in only two species are considered (fig. 1).

In the case of m populations, denoted by $k = 1, \ldots, m$, and each with n_k alleles at a single locus, the equations of change are

$$(p_i^k)' = p_i^k \, w_{i.}^k / \overline{w}^k; \quad (N^k)' = N^k \, \overline{w}^k, \tag{1}$$

$$k = 1, \ldots, m; \; i = 1, \ldots, n_k.$$

Here p_i^k and $(p_i^k)'$ denote respectively the frequencies of the ith allele in the kth population in two successive generations, $\overline{w}^k$ is the mean fitness of the kth population, $w_{i.}^k$ is the mean fitness associated with the ith allele in the kth population, and N^k and $(N^k)'$ are the successive population densities of the kth population (Levin and Udovic, 1977).

As already suggested, it is in general impossible to use models of this form (or multi-locus extensions) as the basis for prediction of the coevolutionary process. Yet they fulfill an indispensable role in providing a basis for thought, and as aids to the understanding of the influences of evolving

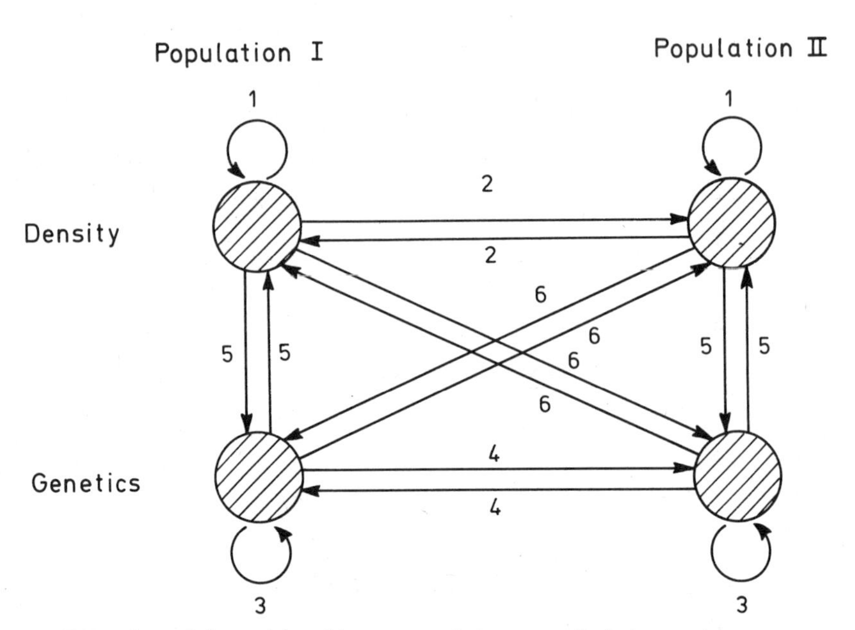

FIGURE 1. Schematic diagram of types of interactions among densities and genetic systems of interacting populations (from Levin and Udovic, 1977). Numbers differentiate different types of interactions.

populations upon one another (Levin and Udovic, 1977; Levin, 1978).

Literal interpretation of models of the form (1) is most nearly justified when they deal with interacting species which are tightly coevolved, for example highly specific mutualist or parasite-host systems (see Janzen, 1980; Feeny, 1982). The best such examples are for the cereal rust gene-for-gene systems already mentioned, in which coevolution is tight and the genetic basis well understood.

Cereal rusts damage their hosts by injecting hyphae to collect nutrients, and forming surface pustules which may contain tens of thousands of uredospores capable of dispersal for thousands of miles and subsequent germination on new host plants (Stakman and Christensen, 1946; Hogg et al., 1969; Van der Plank, 1975). Damage to the host plants may be severe.

Flor (1955, 1956) studied the genetics of the interaction between flax (*Linum usitatissimum*) and the rust *Melampsoa lini*, and his fundamental work has inspired theoretical investigations (e.g., Mode, 1958, 1960, 1961; Person, 1966; Leonard and Czochor, 1978; Lewis, 1978, 1981a, b; Fleming, 1980, 1982) and empirical studies of other cereal-cereal rust associations.

Flor found that there were 27 genes for resistance (R genes) distributed as multiple alleles at five loci in flax. Only two of the loci are linked (with 26 per cent recombination) (Flor, 1955, 1956). Resistance is inherited as a dominant character. Virulence in rust is controlled by a complementary system that identifies in a one-to-one relationship each gene in the host with one in the parasite. Virulence is recessive, and the genes for virulence are located at distinct loci which segregate independently. Because of the "gene-for-gene" duality, resistance will only be operative provided at least one gene for resistance is present in the host *and* the corresponding gene for avirulence is present in the parasite. That is, the (homozygous) virulent parasite can overcome the corresponding resistance in the host.

Such single gene-for-gene relationships have been found in a great many other parasite-host systems (Person, 1966; Pimentel, 1982), although their genericness as models for disease resistance has been sharply challenged (Gracen, 1982). Their structure makes them the perfect subjects for coevolutionary modelling.

Because parasite generation time is much shorter than that in the host, the potential for parasites to evolve virulence to overcome single resistance factors is great. On the other hand, because the primary method of dealing with pathogenicity on cereals is the continual introduction of resistant varieties, the possibility exists for a highly oscillatory relationship between parasite and host. Such oscillations are common in U.S. agricultural cereal systems.

However, in the Middle East Fertile Crescent region, the comparable systems are relatively stable, free of epidemics despite favorable environmental conditions which are paradises of opportunities for plant diseases (Browning, 1974; Dinoor, 1974). Numerous authors (e.g., Browning, 1974, 1980; Pimentel, 1982) argue convincingly that this disparity is a consequence of the genetic oversimplification of American agriculture, and that parasite outbreaks will be rare in systems where host populations have been allowed to evolve heterogeneously or where host individuals with differing resistance mechanisms are interplanted. Host interspersion creates a "flypaper" effect, in which resistant hosts trap individuals which have dispersed from nearby vulnerable plants; this concept has been used in the control of plant pathogens on wheat and oats (Jensen, 1965; Browning and Frey, 1969).

Because of the importance of stability in these systems, models which allow exploration of the conditions under which the parasite-host interactions will stabilize have received considerable study. However, in general these have been closely tied to the gene-for-gene hypothesis and have not considered epidemiological detail or geographical distribution, or aspects such as cross resistance among hosts or host diversity.

Mode (1958) considers a host-parasite model in which there are two resistance genes at a single locus in the host, and two corresponding virulence alleles at independent loci in the pathogen. As with many similar efforts which have built on Mode's approach, the model is formulated entirely in terms of the frequencies of different types within the population, and mass-action assumptions are substituted for a detailed consideration of the host-parasite relationship. Thus the probability that a particular type of host will come into contact with a particular type of pathogen is assumed to be proportional to the frequencies of those types in the population. In later work, Mode (1961) allows for the possibility that some pairings are more likely than others, but according to a fixed preference scheme. Such assumptions, which are more appropriate to predator-prey than to parasite-host interactions, ignore the fact that parasites are non-uniformly distributed over their hosts, especially in microparasitic infections in which the parasite multiplies on the host (Anderson and May, 1979, 1982a). The significance of this will be developed later.

Mode (1958) utilizes a continuous time (weak selection) analogue of the equations of Wright (see Kimura, 1958; Crow and Kimura, 1965), and obtains by standard methods the conditions for stability of polymorphic equilibria. Of course, in the absence of costs to resistance or of constraints (e.g., due to segregation) on its evolution, a host population should evolve towards maximal resistance. Similarly, in the absence of costs and constraints, maximal parasitic virulence should evolve.

But there are costs, some metabolic and others ecological or epidemiological. For example, as resistance at a particular locus declines, parasites which are virulent at that locus are likely to be at a competitive disadvantage. This general feature has been incorporated into most models, including Mode's original work involving many strains of parasites which differ in their loci of virulence (Mode, 1960, 1961). The prototypical model is that introduced by Mode (1960), in which both populations are effectively asexual. As Mode (1960, 1961) discovered, such models never lead to a stable equilibrium unless some form of intraspecific frequency dependence is assumed.

The classical formulation (*a la* Mode) has been explored recently by numerous authors. Leonard (1977) develops a mathematical model which is a slight variant. But as is pointed out by Sedcole (1978), Leonard's model is still unable to support a stable equilibrium. Computer simulations by Leonard and Czochor (1978) suggest the existence of a stable equilibrium. But as Fleming (1980) discusses, this stabilization is due to a transformation of the cereal model from a parallel or simultaneous form to a serial or sequential one (which Leonard and Czochor (1980) argue is biologically more appropriate); this introduces an artificial stabilizing frequency dependence (Levin, 1972; Lewis, 1981a).

All of these models are based on the assumption that fitnesses within each species depend only on the characteristics of the other. Host fitnesses may depend on relative parasite levels, but not on the level of resistance in the host population; parasite fitnesses are determined entirely by the host. The continuous time description of the evolution of a single gene-for-gene system is then

$$\frac{dR}{dt} = R(1 - R)F_R(V) \qquad (2)$$

$$\frac{dV}{dt} = V(1 - V)F_V(R)$$

(Fleming, 1982). Here R is the frequency of the resistant gene in the host, and V is the frequency of virulence in the parasite. $F_R(V)$ and $F_V(R)$ functions of V and R respectively which are often taken to be linear, but that is not essential. The system (2) can never support a stable polymorphic equilibrium; any such equilibrium will be neutrally stable. The orbits of (2) are the solutions to

$$\frac{F_R(V)}{V(1 - V)}\, dV = \frac{F_V(R)}{R(1 - R)}\, dR \qquad (3)$$

and thus have the first integral

$$\int \frac{F_R(V)}{V(1 - V)}\,dV - \int \frac{F_V(R)}{R(1 - R)}\,dR = \text{constant}. \qquad (4)$$

As with the Lotka-Volterra equations, there exist a family of curves (defined by (4)), among which the system can be arbitrarily shifted by perturbations. There is no homeostatic response towards any particular path, no tendency to return to an initial curve after displacement. In general, under reasonable assumptions, these curves will be closed orbits about a neutrally stable equilibrium point (see Fleming, 1982).

If diploid genetics are introduced, it is possible to stabilize such parasite-host interactions by some form of heterozygote non-intermediacy. Mode (1958), in a paper which curiously is not cited in Mode (1960, 1961), considers a diallelic-digenic model of a host-pathogen system. There are three host genotypes -- R_1R_1, R_1R_2, R_2R_2 -- where R_1 and R_2 are distinct alleles for resistance. Pathogenic virulence with respect to R_1 and R_2 is assumed to be controlled at a pair of unlinked loci, although Mode discusses briefly the possible effects of linkage. Mode finds that a stable equilibrium, polymorphic at every locus, can be maintained in this model provided a number of conditions apply. The most critical of these is that a parasite which is heterozygous at a particular locus, when associated with a homozygous host with the corresponding resistance, is at a competitive disadvantage relative to the avirulent homozygote. This assumption means that there is a cost associated with carrying a single recessive gene for virulence. It is further assumed that, on a host homozygous for a given type of resistance, a parasite homozygous for virulence at both loci is at a disadvantage relative to an individual carrying only the necessary virulence. No data are presented to support these assumptions, and in fact they seem contrary to the evidence (Sidhu, 1975). The stable equilibrium in Mode's model involves marginal underdominance in the parasite, and marginal overdominance (cf. Wallace, 1968) in the host.

Fleming (1982) considers a model describing the interaction between an asexual host and an outcrossing parasite, in which the dominance of avirulence is incomplete. He shows that a stable polymorphic equilibrium is possible if the penetration of dominance is greater on susceptible than on resistant hosts. Fleming points out that the resultant stable equilibrium also requires marginal overdominance for the host.

Other attempts to incorporate diploid genetics have treated the parasite as asexual, but the host as a diploid self-fertilizing form. Lewis (1981a, b) develops the general framework, built on the structure and notation (1) developed by Wright (1955; see Levin and Udovic, 1977; Levin, 1978). Related mathematical studies may be found in Jayakar (1970), Levin (1972), Yu (1972), and Clarke (1976).

Following Lewis' treatment, consider the interaction between a panmictic diallelic diploid host and a diallelic haploid pathogen. Let A, a be the host alleles, with frequencies p, $1 - p$; and let B, b be the pathogen alleles, with frequencies q, $1 - q$. In the simplest (symmetric) case, the mean fitnesses of the two pathogen types are given by

$$v_B = p^2\alpha + 2p(1 - p)\beta + (1 - p)^2\gamma \tag{5}$$

and

$$v_b = p^2\gamma + 2p(1 - p)\beta + (1 - p)^2\alpha, \tag{6}$$

in which α, β, and γ are the fitnesses of the pathogens respectively on the host genotypes AA, Aa, aa.

Correspondingly, the host fitnesses in the presence of a

given pathogenic infection are taken to be the complements of the associated pathogenic virulences; thus, for example, $1 - \alpha$ is the fitness of *AA* when associated with pathogen *B*. x is the probability that a given host will be attacked, and uninfected hosts are equally fit. The overall fitnesses of the three host genotypes are then given by

$$\begin{aligned} w_{AA} &= 1 - x + x[q(1 - \alpha) + (1 - q)(1 - \gamma)] \\ &= 1 - x[q\alpha + (1 - q)\gamma] \\ w_{Aa} &= 1 - x + x[q(1 - \beta) + (1 - q)(1 - \beta)] \\ &= 1 - x[\beta] \\ w_{aa} &= 1 - x + x[q(1 - \gamma) + (1 - q)(1 - \alpha)] \\ &= 1 - x[q\gamma + (1 - q)\alpha]. \end{aligned} \tag{7}$$

This model generalizes those of Jayakar (1970) and Yu (1972). However, it is not consistent with those gene-for-gene systems in which resistance is dominant.

Following standard procedure (see Levin, 1978) one obtains the equations of change of gene frequency. Letting primed variables denote frequencies in a successor generation, we obtain

$$p' = p\frac{pw_{AA} + (1 - p)w_{Aa}}{p(pw_{AA} + (1 - p)w_{Aa} + (1 - p)(pw_{Aa} + (1 - p)w_{aa})}$$

$$= p\frac{w_{A.}}{pw_{A.} + (1 - p)w_{a.}} \tag{8}$$

$$q' = q\frac{v_{B.}}{qv_{B.} + (1 - q)v_{b.}} \tag{9}$$

(8) - (9) is a special case of (1), with densities ignored.

As Lewis shows, and as symmetry dictates, the polymorphic equilibrium of this system occurs at $p = q = 0.5$. Stability is analyzed by standard linearization methods, and shown to be tied to the necessary and sufficient condition

$$\beta^2 < \alpha\gamma - \frac{(\alpha - \gamma)^2}{4} = \left(\frac{\alpha + \gamma}{2}\right)^2 - \frac{(\alpha - \gamma)^2}{2}. \qquad (10)$$

Lewis (1981a) acknowledges the validity of Levin and Udovic's (1977) conclusion that in such coevolutionary situations host heterozygote superiority or inferiority is a necessary condition for a polymorphic equilibrium, but argues that it is not clearly interpretable on the individual level. In fact, this claim is a bit misleading. The symmetry of the problem assures that the heterozygote cannot be strictly intermediate at equilibrium; thus the condition is a trivial one. However, something stronger can be said: once again, marginal overdominance is essential. The condition for host heterozygote advantage at equilibrium is

$$\beta < (\alpha + \gamma)/2, \qquad (11)$$

which clearly is a necessary condition for (10) to be satisfied, and hence a necessary condition for stability. Condition (11) is easily interpretable at the individual level. Note that the stricter condition $\beta < \min(\alpha, \gamma)$, which states that the heterozygote is most fit at every pathogen frequency, is neither necessary nor sufficient for the existence of a stable equilibrium.

Lewis notes further that if α and γ are too different, no stable equilibrium is possible, citing the case when one of these vanishes. In fact, it may be easily seen from (10) that the relevant necessary condition is that the right-hand side be positive, which will be the case provided the ratio of the larger to the smaller of α and γ not exceed $3 + 2\sqrt{2}$. If this condition is satisfied, then there are β's satisfying (10); otherwise, there are not. If the ratio is less than 5, $\beta < \min(\alpha, \gamma)$ is a sufficient condition for stability; but in the thin region (of α/γ or γ/α) between 5 and $3 + 2\sqrt{2} \sim 5.8$, heterozygote superiority at all levels still fails to insure the existence of a stable polymorphic equilibrium.

Heterozygote non-intermediacy is in general a necessary condition for the existence of a polymorphic equilibrium in such systems (Levin and Udovic, 1977), although there are degenerate examples in which this condition is not strictly satisfied. For example, in the extreme case that the heterozygote fitness is always the geometric mean of the two homozygote fitnesses, the system (6) - (7) reduces to a doubly haploid model of the type discussed earlier (Nagylaki, 1977; Lewis, 1981b), and creates a rather special situation whereby polymorphic equilibria are feasible with all three fitnesses equal. But as we saw earlier, such models are not inherently stable in character.

In the absence of stabilizing forces, host-parasite systems will have a tendency to oscillate; consequently, one may observe sustained oscillations of limit-cycle or more complicated type, or system collapse. The existence of "invariant circles" -- discrete analogues of limit cycles -- may be studied by way of appropriate versions of the Hopf bifurcation theorem (Guckenheimer et al., 1977) for β close to the threshold suggested by (10). For β larger, it is likely that not only periodic, but aperiodic and chaotic patterns will emerge (see also Anderson and May, 1982a). These may be asymptotically stable, unlike those in the system (2). Study of sustained nonlinear behavior of the solutions to host-parasite equations in the unstable case has been largely restricted to numerical simulation (Yu, 1972; Leonard and Czochor, 1978; Fleming, 1980; Lewis, 1981a).

More general models of host-parasite interactions will lead to stabilization of polymorphic equilibria without heterozygote overdominance. These consider both intraspecific and interspecific frequency and density dependence. There are a wide variety of possible influences on stability, represented by the many feedback loops shown in fig. 1. Following Wright (1955), Levin and Udovic (1977) derive the general set of equations (12), which are a special case of (1):

$$p_i' = p_i w_{i.} / \bar{w} \;,\; q_j' = q_j v_{j.} \;/\; \bar{v} \;,$$
$$N' = N\bar{w} \;,\; M' = M\bar{v} \;. \tag{12}$$

Here, p_i, q_j, N, M denote respectively the gene frequencies and population densities in the two populations. The subscripts distinguish alleles, and hence any number may be treated. The system (8) - (9) is easily seen to be a special case of (12).

System (12) allows consideration of ecological interactions that are more general than host-parasite, and Levin and Udovic (1977) explore the interplay between the "genetical" conditions for stability and the "ecological" ones. Special cases include Levin's (1972) study of stability in host-parasite or prey-predator *genetic feedback* systems, in which host (prey) genetics interact with parasite (predator) densities, and Gillespie's complementary work (Gillespie, 1975), in which intraspecific frequency dependence in host evolution emerges from consideration of parasite numbers. These are discussed in more detail in the next section, as are more general epidemiological models.

Other mechanisms may also serve to stabilize host-parasite systems: environmental patchiness (Karlin and McGregor, 1972; Clarke, 1976), multi-locus effects, inbreeding (see for example Mode, 1961), and alternative hosts (Lewis, 1981b; Fleming, 1982). Some aspects of these also relate to epidemiology, but will not be treated further in this paper.

HOST-PARASITE MODELS INCORPORATING EPIDEMIOLOGICAL CONSIDERATIONS

Anderson and May (1982a) provide a useful categorization of host-parasite models. These include the "explicitly genetic" models of the preceding section; ones which posit some frequency dependence in host genetics; those which "let conventional epidemiological assumptions ... dictate the form of the frequency dependent fitnesses;" and those which focus more on the epidemiology and suppress the genetic details. These represent a spectrum, not a set of discrete choices, and one must let the problem at hand dictate the particular form.

It is nearly dogma in the parasitological literature that host-parasite systems will evolve towards commensalism (Mode, 1958; Burnet and White, 1972; Hoeprich, 1977; Alexander, 1981; Anderson and May, 1982b; B. Levin et al., 1982). As Anderson and May (1982a) summarize, most textbooks simply take as obvious that parasitic species which destroy their hosts cannot long survive, and will be replaced in time by less and less virulent strains in inexorable progress towards commensalism. But the situation is much more complicated.

Evolution in parasite populations represents an interplay between conflicting factors: within an individual host, the race is to the swift and evolution will favor those with the highest rates of reproduction, which is likely to mean those with higher virulence. But the parasite population is a shifting mosaic of demes associated with individual hosts, and the capacity for profligate growth may doom one's host to a shorter life expectancy and reduce the contribution to the larger (mega-) population. Depending on the balance between these factors, some evolution towards attenuation might be expected among parasites, but this attenuation may be checked far short of commensalism (Levin and Pimentel, 1981; Anderson and May,

1982a, Bremermann and Pickering, 1982). These points are most clearly manifest for monoclonal infections, but remain valid even when secondary infections and polyclonality are possible (Levin and Pimentel, 1981; Bremermann and Pickering, 1982). However, theoretical arguments suggest that in polyclonal situations, where the genetic relatedness of individuals associated with a particular host is lessened, parasite evolution towards attenuation should be less effective than if monoclonality obtained. This further suggests that there will be less parasitic evolution towards avirulence in macroparasitic organisms such as helminths, and increased pressure for the development of resistance among their hosts.

One might expect evolutionary pressures in the host to present a less equivocal picture (Person, 1959). In general, there will be no advantage to the host to be susceptible to the ravages of the parasite, and there should be a continual, if slow, evolution of increased resistance. However, as we infer from examples in which the removal of selective pressures results in a loss of resistance, there may be costs associated with the resistance. As suggested earlier, when such costs exist there will be selection against resistance at low disease levels. Thus, as in the case of parasitic virulence, one should expect protected polymorphisms for intermediate levels of resistance. Of course, there will also be cases where resistance becomes fixed and the parasite is virtually eliminated. As long as polymorphism in both species exists, the inherent oscillatory character of the host-parasite system encourages occasional localized outbreaks and even system-wide fluctuations. As already mentioned, such oscillations may be stable and sustained, unlike those observed for the system (2).

Motivated by Haldane's insights (Haldane, 1948), Pimentel (1961, 1968) and Levin (1972) explicitly incorporate cost into a model of the interaction between host (prey) genetics and parasite (predator) density. As discussed in the previous section, Levin (1972) shows that a balanced polymorphism for resistance may result, at levels which will regulate parasite density; but this stabilization requires marginal overdominance. On the other hand, by incorporating epidemiological considerations, Gillespie (1975) was able to obtain stabilization at intermediate levels of resistance in a haploid host, or in a diploid host in which resistance is dominant (as in the gene-for-gene systems) or recessive. Gillespie's work has been extended by Kemper (1982), Longini (1982), and Anderson and May (1982c, d) to include wider classes of disease (see review in Anderson and May, 1982a).

Parasitic evolution towards attenuation is of a different sort, and relies on interdemic selection. Because of the integral nature of the association between host and parasite, and the fact that in monoclonal infections the host is in essence a parasite deme, these systems represent the best of a all possible candidates for interdemic selection to be important. This is, of course, the basis for the conventional wisdom in parasitology that parasite-host systems should evolve reduced virulence. But the fact that there should be a tendency towards some attenuation does not imply that the end result will be commensalism. Available data (e.g., Fenner and Myers, 1978) and theory (Levin and Pimentel, 1981; Anderson and May, 1982a; Bremermann and Pickering, 1982) suggest that stable intermediate levels of resistance are to be expected (see review in B. Levin et al., 1982). For example, the smallpox virus seems to have achieved virulence stasis long ago, and has not undergone substantial change in a millenium and a half (Fenner, pers. comm.).

It has often been pointed out (Lewontin, 1970; Levin and Pimentel, 1981, B. Levin et al., 1982) that landmark studies by Fenner and his colleagues (Fenner and Marshall, 1957;

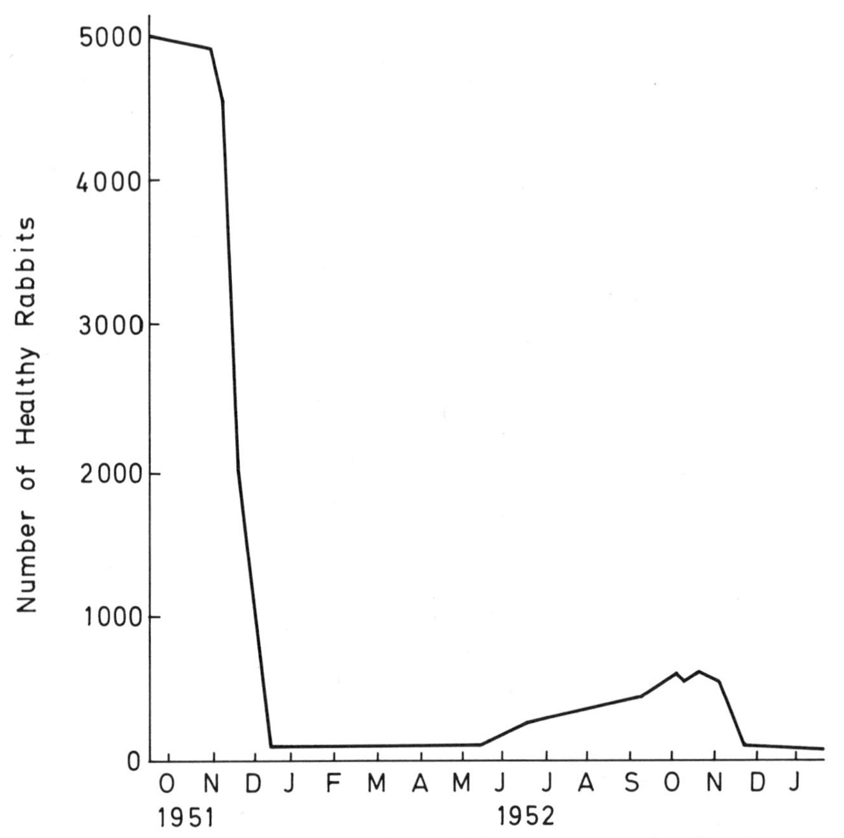

FIGURE 2. Number of healthy rabbits per standardized transect counts at Lake Urana region immediately after the introduction of the myxoma virus into the host rabbit population (after Myers et al., 1954; Levin and Pimentel, 1981).

Fenner, 1965; Fenner and Ratcliffe, 1965; Fenner and Myers, 1978; see also Saunders, 1980) of the interaction between the European rabbit (*Oryctolagus cuniculus*) introduced into Australia and the myxoma virus introduced to control it present perhaps the best documented evidence for the evolution of reduced virulence in host-parasite systems. Similar studies on myxomatosis in *Oryctolagus* in Britain and France are reported in Ross (1982), and discussed by Anderson and May (1982a).

In the Australian system, *Oryctolagus* was at outbreak

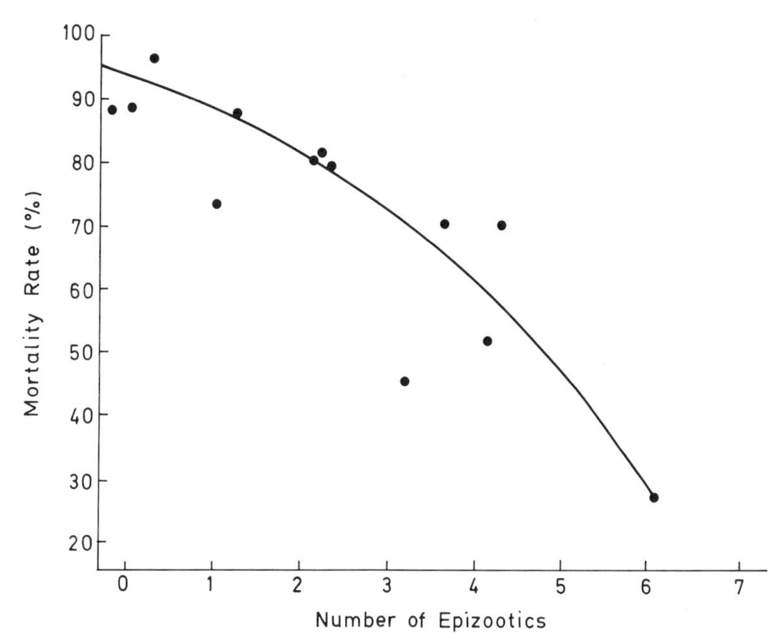

FIGURE 3. Mortality rates of wild rabbits from Lake Urana region after exposures to several epizootics of myxoma virus, after challenge infection with strain of myxoma virus grade III virulence (after Fenner and Myers, 1978; Levin and Pimentel, 1981). Abscissa is weighted for immune rates among survivors of each epizootic.

levels by 1880. The myxoma virus, which lives in the South American tropical rabbit *Sylvilagus brasiliensis* without producing any generalized disease, was introduced in 1950 as a control mechanism, and *Oryctolagus* showed a 99.8% case mortality. This led to a near elimination of the rabbit population (fig. 2). Evolution in the virus towards reduced virulence was rapid, and was subsequently reinforced by evolution of resistance in the rabbit. This led to an overall reduction in observed mortality (fig. 3). Whereas the viral population in 1950-51 was composed almost totally of grade I virulence individuals, by 1963-64 grades I and II were virtually eliminated and the average individual was about a IIb

Table 1. Virulence of Field Myxoma Virus Types in Rabbits in Australia (after Fenner and Myers, 1978).

	VIRULENCE TYPE GRADE					
Grade of Severity	I	II	IIIA	IIIB	IV	V
Mean survival times of rabbits (days)	<13	14-16	17-22	23-28	29-50	...
Case-mortality rate (%)	>99	95-99	90-95	70-90	50-70	<50
Australia						
1950-1951	100	...	...	...	...	...
1958-1959	0	25	29	27	14	5
1963-1964	0	.3	26	33	31	9

(see table 1). Similar trends were observed in Great Britain and France (Anderson and May, 1982a; Ross, 1982) after introduction of myxoma in 1952. It is not yet known whether these represent stable equilibria or whether there will be a continuing trend towards avirulence.

Inspired by Fenner's work, Levin and Pimentel (1981) analyze evolution of avirulence in parasites by means of a model based on the system described in fig. 4. The fundamental nature of the model is that hosts are divided into a number of categories based on which parasites they harbor; only two strains are considered, but in principle any number could be. The model ignores parasite densities on a host, and considers only whether the host has been infected with a particular strain. This is an obvious oversimplification, but

---

FIGURE 4. Schematic diagram of transitions and rates in interaction between host and two strains of virulence.

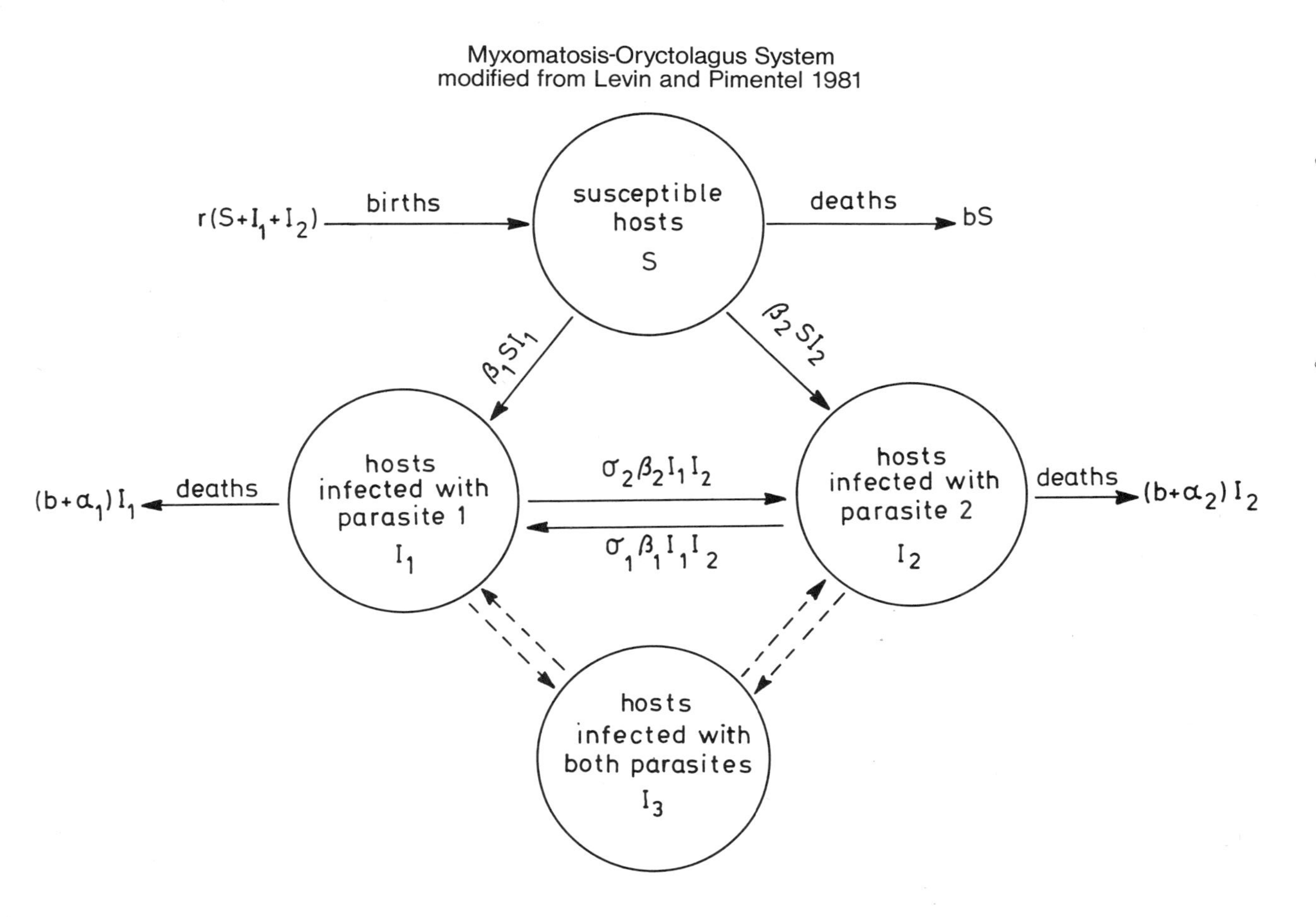
Myxomatosis-Oryctolagus System
modified from Levin and Pimentel 1981
$r(S+I_1+I_2)$
births
susceptible hosts
S
deaths
bS
$\beta_1 SI_1$
$\beta_2 SI_2$
$(b+\alpha_1)I_1$
deaths
hosts infected with parasite 1
I_1
$\sigma_2\beta_2 I_1 I_2$
$\sigma_1\beta_1 I_1 I_2$
hosts infected with parasite 2
I_2
deaths
$(b+\alpha_2)I_2$
hosts infected with both parasites
I_3

does not affect the main conclusions. Nonetheless, given the relation of mortality to parasite load (Anderson and May, 1981) and the influence on transmission characteristics, it would be of interest to examine the effects of parasite densities.

Based on fig. 4, one obtains immediately the system

$$\frac{dS}{dt} = r(S + I_1 + I_2) - bS - \beta_1 SI_1 - \beta_2 SI_2$$

$$\frac{dI_1}{dt} = \beta_1 SI_1 - (b + \alpha_1)I_1 + (\sigma_1\beta_1 - \sigma_2\beta_2)I_1I_2 \tag{13}$$

$$\frac{dI_2}{dt} = \beta_2 SI_2 - (b + \alpha_2)I_2 + (\sigma_2\beta_2 - \sigma_1\beta_1)I_1I_2$$

in which S is the number of susceptible rabbits, I_1 is the number of rabbits infected with parasite 1, and I_2 is the number of rabbits infected with parasite 2. All hosts have the same birth rate r, and newborns are uninfected. b is the death rate of susceptible individuals, and $b + \alpha_i$ is the death rate of those housing parasite i. β_i is the transmissibility of strain i, and σ_i measures the susceptibility to invasion by strain i of individuals infected with the competitor strain. Here invasion implies competitive displacement of the resident parasite. For simplicity, the mixed stage I_3 is ignored. This assumption will be relaxed later (system (20)), but consideration of the intermediate stage does not affect qualitatively the results. I shall also later relax the assumptions that the birth rates for all hosts are identical, and will allow for recovery from infection.

Parasite 2 is the virulent strain, and hence $\alpha_2 > \alpha_1$. In Levin and Pimentel (1981), it is assumed that $\beta_1 = \beta_2$ and $\sigma_1 = 0$. However, as observed by Anderson and May (1982a), the model and its conclusions apply equally when $\sigma_1 \neq 0$. Then, assuming $r > b$, the system (13) with $\beta_1 = \beta_2$ supports a locally stable polymorphism provided the relativized life expectancy of the virulent-infected host, $r/(b + \alpha_2)$, is sufficiently short, and that of the avirulent-infected host, $r/(b + \alpha_1)$, is sufficiently long. Specifically, the conditions are

$$\frac{r}{b + \alpha_2} < 1 - \frac{(\sigma_2 - \sigma_1)(r - b)}{\alpha_2 - \alpha_1} < \frac{r}{b + \alpha_1} \tag{14}$$

If the first inequality in (14) is violated, the virulent strain will eliminate the avirulent; if the second is violated, avirulence will win. But if (14) is satisfied, intermediate levels will result, at the levels

$$\overline{S} = r/\theta, \quad \overline{I}_1 = \frac{b + \alpha_2 - \beta\overline{S}}{\beta(\sigma_2 - \sigma_1)}, \quad \overline{I}_2 = \frac{\beta\overline{S} - (b + \alpha_1)}{\beta(\sigma_2 - \sigma_1)}, \tag{15}$$

where

$$\theta = \beta(1 - (\sigma_2 - \sigma_1)(r - b)/(\alpha_2 - \alpha_1)). \tag{16}$$

These conditions are simply a translation of the results obtained by Levin and Pimentel (1981). Coexistence of strains here depends on the invasion of avirulent-infected hosts by virulent viruses. If the right inequality in (14) is violated, a new steady state is attained at host densities too low to allow the virulent strain to survive in the population.

It is important to note that selection for intermediate levels of virulence does not depend on coexistence between strains. The discussion so far simply indicates that under some conditions a less virulent strain will displace a virulent one; under others the reverse will be true; and under others there will be coexistence. A generalization of the results given here would demonstrate that when many strains are placed in competition, the outcome may be coexistence, competitive dominance by a single intermediate type, or selection for an extreme type.

More generally, assume $\beta_1 \leq \beta_2$ (the transmissibility of the virulent strain is at least equal to that of the avirulent strain). Then, provided $\sigma_1\beta_1 < \sigma_2\beta_2$, the polymorphic equilibrium (15) generalizes to

$$\bar{S} = r/\theta^*, \quad \bar{I}_1 = \frac{b + \alpha_2 - \beta_2\bar{S}}{\beta_2\sigma_2 - \beta_1\sigma_1}, \quad \bar{I}_2 = \frac{\beta_1\bar{S} - (b + \alpha_1)}{\beta_2\sigma_2 - \beta_1\sigma_1}, \tag{17}$$

where

$$\theta^* = \frac{\alpha_2\beta_1 - \alpha_1\beta_2}{\alpha_2 - \alpha_1} + \frac{(\sigma_1 - 1)\beta_1 - (\sigma_2 - 2)\beta_2}{\alpha_2 - \alpha_1}(r - b). \tag{18}$$

As before, the local stability condition is that (17) be feasible; i.e.

$$\beta_2 \frac{r}{b + \alpha_2} < \theta^* < \beta_1 \frac{r}{b + \alpha_1}, \tag{19}$$

which reduces to (14) if $\beta_1 = \beta_2$.

In interpreting (19), it should first be noted that $\beta r/(b + \alpha)$ is a measure of the transmission rate of a strain, the product of its transmissibility times the relativized expected lifetime of its host. θ^* measures the relative strengths of intra-demic selection and inter-demic selection, where a deme is the total parasite colony associated with a particular host.

For many diseases, secondary infections may be uncommon, especially if the course of the disease is swift. Nonetheless, we expect some susceptibility, including cases where the host has fought off the primary infection. σ is a measure of the relative susceptibility of an already infected host; normally it will be between 0 and 1 because $\sigma\beta$ measures not the rate of invasion by a secondary infection, but the rate at which a secondary infection becomes the dominant one. For such considerations, the model (13) is an oversimplification, and it is best to consider explicitly the intermediate stage I_3 (both infections present) and the appropriate rates of recovery.

The equations of interest then become

$$\frac{dS}{dt} = (r_0S + r_1I_1 + r_2I_2 + r_3I_3) - bS - \beta_1SI_1 - \beta_2SI_2 + v_1I_1 + v_2I_2$$

$$\frac{dI_1}{dt} = \beta_1SI_1 - (b + \alpha_1)I_1 - v_1I_1 + w_2I_3 - \gamma_2\beta_2I_1I_2$$

$$\frac{dI_2}{dt} = \beta_2SI_2 - (b + \alpha_2)I_2 - v_2I_2 + w_1I_3 - \gamma_1\beta_1I_1I_2 \quad (20)$$

$$\frac{dI_3}{dt} = (\gamma_1\beta_1 + \gamma_2\beta_2)I_1I_2 - (w_1 + w_2)I_3 - (b + \alpha_3)I_3 .$$

Here, (13) has been generalized in several ways. First, infected hosts are allowed to have a different reproductive rate from uninfected, the deviation being dependent on the nature of the infection. Second, v_i is the rate at which hosts infected with strain i alone recover, and w_i is the rate at which those infected with both strains lose strain i. Third, $b + \alpha_3$ is the death rate of doubly infected hosts. Finally, $\gamma_i\beta_i$ is the rate at which hosts infected with strain $j \neq i$ acquire a secondary infection by i. γ_i is not identical with σ_i in (13), since it measures only the rate of acquisition of the secondary infection, not the rate of displacement. Thus $\gamma_i \geq \sigma_i$.

If $w_1 + w_2 >> (b + \alpha_3)$-- that is, if the rate of intrahost competitive exclusion is fast relative to the death rate of doubly infected hosts -- and if the dynamics of I_3 are taken to be fast, then a pseudo-steady-state assumption allows the approximation

$$I_3 \cong (\gamma_1\beta_1 + \gamma_2\beta_2)I_1I_2 / (w_1 + w_2) \quad (21)$$

to be made in (20). This reduces the equation for dI_1/dt, for example, to

$$\frac{dI_1}{dt} = \beta_1SI - (b + \alpha_1 + v_1)I_1 + (\sigma_1\beta_1 - \sigma_2\beta_2)I_1I_2 , \quad (22)$$

where

$$\sigma_1 = \gamma_1 \frac{w_2}{w_1 + w_2} \text{ and } \sigma_2 = \gamma_2 \frac{w_1}{w_1 + w_2}. \tag{23}$$

Thus the relationship between σ_i and γ_i is made clearer.

More generally, if the pseudo-steady-state approximation is not made, the system (20) may still possess a stable polymorphic equilibrium. The conditions are complicated (Levin, 1983), and are not given in detail here. However, most important is that for a stable polymorphic equilibrium to exist, necessarily the transmission rates satisfy a condition

$$\beta_2 r_0 \,/\, (b + \alpha_2 + v_2) < \theta^{**} < \beta_1 r_0 \,/\, (b + \alpha_1 + v_1), \tag{24}$$

where θ^{**} is a generalization of (18), and further

$$\gamma_1 \beta_1 w_2 > \gamma_2 \beta_2 (w_1 + b + \alpha_3) \tag{25}$$

(which replaces the earlier condition $\sigma_2\beta_2 > \sigma_1\beta_1$).

If secondary infections are ignored, but recovery is permitted, (20) reduces to

$$\frac{dS}{dt} = (r_0 S + r_1 I_1 + r_2 I_2) - bS - \beta_1 S I_1 - \beta_2 S I_2 + v_1 I_1 + v_2 I_2$$

$$\frac{dI_1}{dt} = \beta_1 S I_1 - (b + \alpha_1 + v_1) I_1 \tag{26}$$

$$\frac{dI_2}{dt} = \beta_2 S I_2 - (b + \alpha_2 + v_2) I_2.$$

Then, as Anderson and May (1982a, c) and Bremermann (1980, Bremermann and Pickering, 1982) discuss for the case $r_0 = r_1 = r_2$, coexistence is not possible; the winning strain is that which maximizes the ratio $\beta/(b + \alpha + v)$, subject to whatever constraints exist. Note that this is equivalent to maximizing the intrinsic reproductive rate

$$R_0 = \frac{\beta}{b + \alpha + v} N \tag{27}$$

of the parasite (Dietz, 1975) for a given population density N of hosts. Thus it defines an evolutionarily stable strategy (Maynard Smith, 1976, 1977).

Anderson and May (1982a) derive a functional relationship between α and v and maximize R_0 (for N fixed); b is determined from data, and β is taken to be a constant. A similar technique was suggested independently by Bremermann and Pickering (1982). Using data taken from Fenner and Ratcliffe (1965) to determine the parameters and the functional relationships, Anderson and May (1982a) determine a theoretical optimal type, which is remarkably close (but not identical) to the modal strain which emerged in the Australian system; similar conclusions apply to the European data. Of course, as stated earlier, it is not yet certain whether either the Australian or European system has stabilized.

In conclusion, models which take into account the details of the epidemiological distribution of populations can demonstrate behavior not possible in the more classical host-parasite models. In this section I have discussed some of the most primitive of such models, but these are sufficient to provide insights into the evolution of avirulence in host-parasite associations. Because in natural populations parasite burdens are highly heterogeneous in their distributions, it would be worthwhile to examine extended models which incorporate more detail regarding population distributions.

Epidemiological considerations allow one to recognize that populations have a demic structure, with restricted flow between demes; it is well recognized that such structure is important for selection to favor group-oriented behavior (D. Wilson, 1977; Wade, 1978). Demic structure is simply a special case of geographical structure, which is also known to be important in any evolutionary analysis (Malécot, 1948; Wright, 1949; Kimura and Weiss, 1964; Maruyama, 1971; Karlin and

McGregor, 1972; Nagylaki, 1978) as well as in ecological considerations (Levin, 1974, 1976, 1981). Many traits, for example allelopathy, are locally specific in their nature, and their spread is determined by the geographical structure of the environment. In an elegant recent paper (Chao and B. Levin, 1981) , it is shown that *E. coli* that produce an anti-competitor toxin (colicin) may be favored by selection when rare, but only in a structured habitat, a soft agar matrix. This is because the costs of producing the toxin reduce the intrinsic rate of increase of the colicinogenic bacteria, which places them at a competitive disadvantage in liquid cultures which are well-mixed. In agar, clones form inhibition zones around themselves, and these permit their spread.

DIFFUSE COEVOLUTION

Although the general formulation (1) is unrestricted in dimension, the models so far considered deal principally with tight coevolution between a pair of closely associated species. But many problems of interest involve diffuse coevolution involving many species. For example, immune systems represent a generalized response of vertebrates to a suite of possible hostile agents, and it would be fruitless to try to study such responses in models of the form of the previous chapters. Similar remarks apply to the evolution of chemical defenses in plants, which are to a large extent also a generalized response to diverse enemies (Feeny, 1975, 1982; Janzen, 1980). Clarke (1975, 1976) has focused attention on the influence of natural enemies in the maintenance of diversity, and others have addressed more specifically the importance of natural enemies to the evolution of sex as a diversifying mechanism (Jaenike, 1978; Hamilton, 1980, 1982). To consider such questions, which deal with the coevolution of many factors on a virtual continuum of responses, requires a new point of view.

The problem of the evolution of a diversity of anti-enemy defense mechanisms is a restriction of the more general problem of the evolution of aspect diversity (Rand, 1967; Ricklefs and O'Rourke, 1975; Endler, 1978; Levin and Segel, 1982), with the notion of aspect interpreted broadly. To approach this problem, Lee Segel and I (Levin and Segel, 1982) have proposed a quite different framework which builds on models of the evolution of quantitative characters, such as those discussed by Slatkin (1970) and others (Bossert, 1963; Kimura, 1965), coupled with descriptions of ecological dynamics. Related models have been utilized by Roughgarden (1972) and Rocklin and Oster (1976) in studying the evolution of competition. The approach is introduced briefly in this section; for the details, the reader is referred to Levin and Segel (1982). As discussed more generally earlier, such models are not intended to be used for literal prediction, but rather to help in understanding the evolutionary patterns which have emerged, to aid in framing questions, and to identify those key parameters which seem to control the dynamic responses.

The approach is illustrated most easily for predator-prey interactions, for they permit a mass-action formulation. As discussed in the previous section, for parasite-host interactions it is important to revise classical approaches to consider epidemiology, and that represents an important future extension of the work discussed here. Another extension is concerned with plant-pollinator associations, in which many of the same questions are at issue (Schemske, this volume): What accounts for diversity? What determines the specificity of associations? That is, under what conditions will the community evolve towards generalized associations, as opposed to rather specific (predator-prey, parasite-host, pollinator-plant) assemblages? Timm's discussion (this volume) of conditions in parasite-host communities which favor resource tracking versus Farenholz's rule highlights this dichotomy.

In the model presented here, the generation time of the predators is considered to be longer than that of the prey, so that coevolution among the various prey species is taking place on the same time scale as that of the ecological responses of the prey. Evolution in the predators is not treated, although it would be a straightforward extension of the model to incorporate it.

It has often been suggested (Cain and Sheppard, 1950, 1954; Rand, 1967; Clarke, 1969; Ricklefs and O'Rourke, 1975) that apostatic selection, in which predators preferentially form a labile search image for more common prey, is an important mechanism in determining the diversity of aspects (color, wing or shell patterns, defensive characteristics, etc.) in prey. Although the significance of apostasis in maintaining aspect diversity is debatable, predation in general is of obvious importance (Endler, 1978). Models can help us to assess the relative importance of the various mechanisms.

The model summarized here assumes that prey aspect may be arrayed along a single gradient, scored by z. This single-dimensional illustration is chosen for ease of presentation, and represents no fundamental restriction. Prey aspect is assumed to be genetically determined; predator search image for a particular aspect shifts in inverse relation to the numbers of available prey of that aspect. Prey may mate more readily with similar types (assortative mating) or preferred types (sexual selection). The prey are assumed to be distributed along the aspect gradient z according to the density function $v(z, t)$, which changes over evolutionary time. Similarly, predator search image is distributed at time t according to a density function $e(z, t)$, which shifts as predators shift search images.

We have proposed using this scheme to examine both intraspecific evolution, as for example the diversification of spot patterns in guppies (*Poecilia reticulata* Peters) discussed by Endler (1978), and interspecific coevolution in high diversity communities, as for example the proliferation of wing patterns among moths (Ricklefs and O'Rourke, 1975). Ecologically, we believe that the mechanisms are the same. Of course, there are fundamental differences in the genetic constraints which must be imposed at the intraspecific versus the interspecific level, especially regarding hybridization and outcrossing. But the extremes of random mating and complete reproductive isolation simply represent the poles of a spectrum of possibilities observed in Nature, and the biological species concept is based on a separation of this spectrum. The art of model building at any level of description requires the suppression of much detail at lower levels, and the summary of this detail in a few macroscopic parameters. Thus, to a first approximation we suggest that mating structure can be represented by the assortativity of the mating function, with the interspecific case corresponding to total or near-total positive assortativity.

Rare prey are relatively favored by predator apostasis, but to what degree? If predator search image is broad, then prey with aspects close to target prey benefit less than otherwise from the predator's attention to other aspects. Similarly, but more subtly, positive assortative mating which extends to nearby types can result in the depression of their growth rates and serve as a strong influence in population subdivision and the formation of reproductive isolation.

What are the relative importances of the competing tendencies to subdivision and diversification? When will uniform distributions result, in which all types are equally represented? When will peaks occur (precursors to population

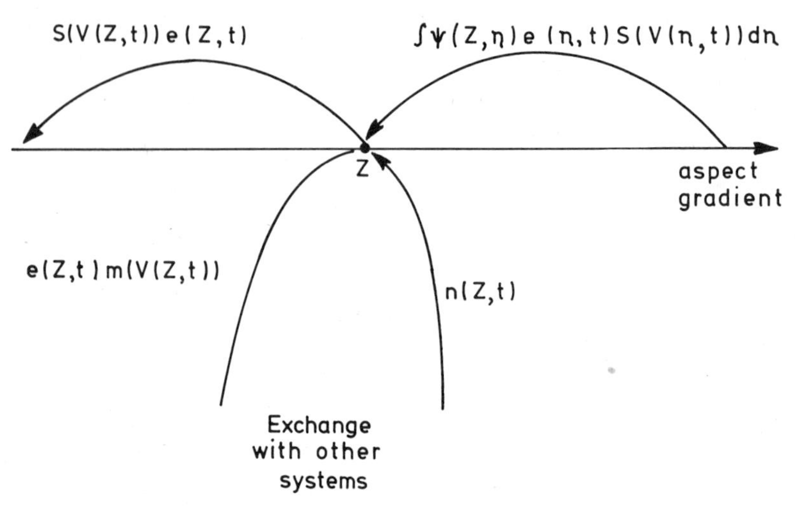

FIGURE 5. Schematic diagram of (rates of) transitions of predator search image. Symbols explained in text.

subdivision and possibly speciation), in which a restricted number of types are numerically dominant, and in which predator-prey associations will become relatively specialized (with predators locked in ecological time to the prey's genetically fixed aspects)? What determines the diversity of types and their relative similarities? When can more general patterns of diversity result, such as the persistent fluctuations analogous to those observed in many parasite-host assemblages?

Predator switching is governed by the transitions shown in fig. 5. The possibilities exist both for switching to prey of a different aspect within the evolutionary group being considered, and for switching to a different class of prey altogether. The latter, described by the function m, is allowed to depend not only on the local density $v(z, t)$, of prey but more generally on some weighted average $V(z, t)$ of prey densities representing the predator's recent experiences and range of selectivities. m is related inversely to V.

Similarly, predator density is measured by a function $e(z, t)$. Predators shift among the available aspects according to a switching function s (which bears an inverse relationship to v or to some weighted average of v), and a redistribution "kernel" $\Psi(z, \eta)$ (which is the distribution of the probability that a predator switching from aspect z will select aspect η as its new search image). Most typically, but not necessarily, $\Psi(z, \eta)$ is a unimodal function with peak at z. The function $n(z, t)$ includes new recruits plus the reentry of predators from alternative prey (outside the evolving group being considered). More generally, n could be allowed to depend upon predator diversity in some way, although on the time scale of interest we expect such dependency to be weak -- recall that predator search image is assumed to have no genetic component. θ defines the weight function implicit in V. Thus, summarizing the exchanges shown in fig. 5, we obtain the equation

$$\frac{\partial e}{\partial t} = - m(V_1)e + n(e, v) - s(V_2)e + \int \Psi(z, \eta)s(V_2)e\,d\eta, \quad (28)$$

where s has been allowed to depend on the weighted average V_2 rather than simply on v, and V_1 is the weighted average defined by θ.

Prey evolution is governed by a model similar to that discussed by Slatkin (1970) for polygenic characters, supplemented by a predator per capita consumption rate $f(v)$. The approach of Slatkin is modified to include a continuous time description, most appropriate if selection is weak, and to allow for density- and frequency-dependent influences. The relative availability of desirable mates for prey of a particular aspect z is measured by the weighted average.

$$W(\eta, t) = \int \alpha(\eta, \xi)v(\xi, t)d\xi \ , \quad (29)$$

in which $\alpha(\eta, \xi)$ defines a preference index. The per capita number of offspring per female, r, depends on W and perhaps on η. Finally, the function $\Phi(\eta, \xi; z)$ is the redistribution kernel, describing the probability distribution for the offspring of a mating between an η-female and a ξ-male. These definitions lead to the equation (30)

$$\frac{\partial v}{\partial t} = -f(v)e + \iint v(\eta, t)v(\xi, t)\alpha(\eta, \xi) \frac{r(\eta, W(\eta, t))}{W(\eta, t)} \Phi(\eta, \xi; z)d\xi d\eta$$

for the rate of change of the victim aspect distribution with time.

The behavior of the system (28) - (30), under particular assumptions, is discussed in Levin and Segel (1982). All forms of patterns of diversity alluded to earlier are observed: uniform distributions, multimodal distributions representing population subdivision, and oscillatory distributions. These, and their descriptive parameters, are related to such critical variables as the switching rates and sensitivities of the predators, predation pressure, selectivity of mating, and the characteristics of the offspring distribution of a given mating. Some analytical results are posssible; other conclusions can emerge only from numerical treatment.

The significance of the example presented is not in the detailed conclusions which may be drawn from it, but as a framework fundamentally different from the ones appropriate for the study of tight coevolution. The model just described may be extended to other situations, e.g. plant-pollinator assemblages. The point is that when one is concerned with general patterns at the community level -- degree of specialization, diversity, etc. -- the most appropriate models are ones which emerge from knowledge of the population genetic basis of inheritance, but which do not retain superfluous genetic detail. Of course, deciding what is superfluous is not always easy, and must be determined by the intuitions and arrogance of the investigator.

SUMMARY

Coevolutionary considerations are essential in understanding evolution in natural contexts, and introduce difficulties which extend those of frequency-dependent responses in classical evolutionary theory. Coevolution implies nonlinear feedback among interacting species, and such nonlinearities substantially complicate any attempt to understand evolutionary change.

The classical approach, based on explicit description of gene frequency change, is most appropriate to small ensembles of tightly interacting species in which the genetic basis of change is well understood. The gene-for-gene resistance systems of cereal plants and their associated pests provide the ideal systems for such modelling, and in this paper the relevant literature is briefly reviewed. Models of quantitative inheritance are not discussed in this context, but are appropriate for tight coevolution when the genetic basis is polygenic.

More generally, because of the tight relationship between parasite and host, the parasite-host interaction is an ideal one in which to study pairwise coevolution (Day, 1974). One of the most enticing consequences of the association is the evolution of reduced virulence in parasites in order to preserve their hosts (Fenner, 1965; Lewontin, 1970). However, to study such evolution requires an approach which deviates from the classical mold and explicitly incorporates epidemiological considerations (Levin and Pimentel, 1981; Anderson and May, 1982a; Bremermann and Pickering, 1982). Such models are discussed in section IV, and hold tremendous potential for the examination of the evolution of host-parasite systems. Anderson and May (1979, 1981; May and Anderson, 1979), by their introduction of host-parasite dynamics into classical parasitology and epidemiology, have already had a major influence on the development of those fields (Anderson and May, 1982b), and all of the work discussed in section IV builds on their approach. Thus, it is all of recent vintage and preliminary. Extensions will certainly broaden our understanding of the dynamics of these systems.

Finally, it is argued that diffuse coevolution (sensu Janzen, 1980; Feeny, 1982) requires yet a different perspective, particularly when attention is focused on system-level integrative parameters such as diversity or degree of specialization. An approach due to Levin and Segel (1982), but building on previous work of others on quantitative inheritance, is outlined and some specifics discussed. This too represents a new paradigm, one which is likely to be a productive source of insights into diffuse coevolution.

No single approach holds all of the answers for the examination of coevolutionary processes. Classical population genetics has not often addressed itself to ecological interactions, whereas ecologists have fallen back too comfortably on adaptationism and optimalogy without adequate recognition of the inherent contradictions in that approach (Lewontin, 1977; Levin, 1981). More recent applications of game theory and the development of the concept of evolutionarily stable strategies (Maynard Smith, 1976, 1977) are advances; but they still retain a number of difficulties, including especially the fact that they are equilibrial in nature. Even when equilibrial approaches are justified, they sometimes provide little information about processes. In this paper, I have emphasized the importance of dynamic models, ranging from detailed genetic ones to ones couched entirely in phenotypic terms. All of these approaches contribute to our understanding of evolution.

ACKNOWLEDGEMENTS

I am pleased to acknowledge the support of the National Science Foundation under grant MCS80-01618 and the Environmental Protection Agency, through its support of the Ecosystems Research Center, Cornell, under grant CR807856. The US-Israel Binational Foundation supported the collaborative work with Lee A. Segel. I am especially grateful to Mark Harwell and Richard A. Fleming for detailed comments on an earlier draft of the manuscript.

LITERATURE CITED

ALEXANDER, M. 1981. Why microbial predators and parasites do not eliminate their prey and hosts. Annual Review of Microbiology, 35:113-133.

ANDERSON, R.M., and R.M. MAY. 1979. Population biology of infectious diseases: Part I. Nature, 280:361-367.

ANDERSON, R.M., and R.M. MAY. 1981. The population dynamics of microparasites and their invertebrate hosts. Philosophical Transactions of the Royal Society of London, Series B 291:451-524.

ANDERSON, R.M., and R.M. MAY. 1982a. Coevolution of hosts and parasites. Parasitology. *To appear*.

ANDERSON, R.M., and R.M. MAY (editors). 1982b. Population Biology of Infectious Diseases. Dahlem Konferenzen. Springer-Verlag, Heidelberg. *To appear*.

ANDERSON, R.M., and R.M. MAY. 1982c. Frequency and density dependent effects in the coevolution of hosts and parasites. *Manuscript*.

ANDERSON, R.M., and R.M. MAY. 1982d. Directly transmitted infectious diseases: control by vaccination. Science, 215:1053-1060.

BOSSERT, W.H. 1963. Simulation of character displacement. Unpublished Doctoral Dissertation, Harvard University, Cambridge, Massachusetts.

BREMERMANN, H.J. 1980. Sex and polymorphism as strategies in host-pathogen interactions. Journal of Theoretical Biology,

87:671-702.

BREMERMANN, H.J., and J. PICKERING. 1982. A game theoretical model of parasite virulence. *Submitted.*

BROWNING, J.A. 1974. Relevance of knowledge about natural ecosystems to development of pest management programs for ecosystems. Proceedings of the American Phytopathological Society, 1:191-199.

BROWNING, J.A. 1980. Genetic protective mechanisms of plant-pathogen interactions: their coevolution and use in breeding for resistance; pp. 52-75. *In* Harris, M.K. (ed.), Biology and Breeding for Resistance to Arthropods and Pathogens in Agricultural Plants. Texas A & M Press, College Station, Texas.

BROWNING, J.A., and K.J. FREY. 1969. Multiline cultivars as a means of disease control. Annual Review of Phytopathology, 7:355-382.

BURNET, M., and D.O. WHITE. 1972. Natural History of Infectious Disease. Cambridge University Press, Cambridge.

CAIN, A.J., and P.M. SHEPPARD. 1950. Selection in the polymorphic land snail *Cepaea nemoralis*. Heredity, 4:275-294.

CAIN, A.J., and P.M. SHEPPARD. 1954. Natural selection in *Cepaea*. Genetics, 39:89-116.

CHAO, L., and B.R. LEVIN. 1981. Structured habitats and the evolution of anticompetitor toxins in bacteria. Proceedings of the National Academy of Sciences, USA, 78:6324-6328.

CLARKE, B.C. 1969. The evidence for apostatic selection. Heredity, 24:347-352.

CLARKE, B.C. 1975. The causes of biological diversity. Scientific American, 233(2):50-60.

CLARKE, B.C. 1976. The ecological genetics of host-parasite relationships; pp. 87-103. *In* Taylor, A.E.R., and R. Muller (eds.), Genetic Aspects of Host-Parasite Relationships. Blackwell, Oxford.

CROW, J.F., and M. KIMURA. 1965. Evolution in sexual and asexual populations. American Naturalist, 103:89-91.

DAY, P.R. 1974. Genetics of Host-Parasite Interactions. W.H. Freeman, San Francisco.

DIETZ, K. 1975. Transmission and control of arbovirus diseases; pp. 104-121. *In* Ludwig, D., and K.L. Cooke (eds.), Epidemiology. Society of Industrial and Applied Mathematics, Philadelphia.

DINOOR, A. 1974. The role of the alternate host in amplifying the pathogenic variability of oat crown rust. Research Reports on Science and Agriculture, Hebrew University of Jerusalem, 1:734-735.

ENDLER, J.A. 1978. A predator's view of animal color patterns. Evolutionary Biology, 11:319-364.

FEENY, P. 1975. Biochemical coevolution between plants and their insect herbivores; pp. 3-19. *In* Gilbert, L.E., and P. H. Raven (eds.), Coevolution of Animals and Plants. University of Texas Press, Austin and London.

FEENY, P. 1982. Coevolution of plants and insects. Chapter 11. *In* Odhiambo, T.R. (ed.), Current Themes in Tropical Sciences, 2: Natural Products for Innovative Pest Management. Pergamon Press, Oxford.

FENNER, F. 1965. Myxoma virus and *Oryctolagus cuniculus*: two colonizing species; pp. 485-501. *In* Baker, H.G., and G.L. Stebbins (eds.), The Genetics of Colonizing Species. Academic Press, New York.

FENNER, F., and I.D. MARSHALL. 1957. A comparison of the virulence for European rabbits (*Oryctolagus cuniculus*) of strains of myxoma virus recovered in the field in Australia, Europe and America. Journal of Hygiene, 55:149-191.

FENNER, F., and K. MYERS. 1978. Myxoma virus and myxomatosis in retrospect: the first quarter century of a new disease; pp. 539-570. *In* Kurstak, E., and K. Maramorosch (eds.), Viruses and Environment. Third International Conference on Comparative Virology, Mont Gabriel, Quebec.

FENNER, F., and F.N. RATCLIFFE. 1965. Myxomatosis. Cambridge University Press, Cambridge.

FLEMING, R. 1980. Selection pressures and plant pathogens: robustness of the model. Phytopathology, 70:175-178. (Errata 1981, 71:268).

FLEMING, R. 1982. Stability properties of simple gene-for-gene relationships. *To appear.*

FLOR, H.H. 1955. Host-parasite interaction in flax rust--its genetics and other implications. Phytopathology, 45: 680-685.

FLOR, H.H. 1956. The complementary genic systems in flax and flax rust. Advances in Genetics, 29-54. Academic Press, New York.

GILLESPIE, J.N. 1975. Natural selection for resistance to epidemics. Ecology, 56:493-495.

GOULD, S.J. 1977. Ever Since Darwin. Norton, New York.

GRACEN, V. 1982. Role of genetics in etiological phytopathology. Annual Review of Phytopathology, 20:219-233.

GUCKENHEIMER, J., G. OSTER, and A. IPAKTCHI. 1977. Dynamics of density-dependent population models. Theoretical Population Biology, 4:101-147.

HALDANE, J.B.S. 1948. The theory of a cline. Journal of Genetics, 48:277-284.

HAMILTON, W.D. 1980. Sex versus non-sex versus parasite. Oikos, 35:282-290. Reprinted in R.W. Hiorns and D. Cooke (editors), The Mathematical Theory of the Dynamics of Biological Populations II, pp. 139-155. Academic Press, London.

HAMILTON, W.D. 1982. Pathogens as causes of genetic diversity in their host populations. *In* Anderson, R.M., and R.M. May (eds.), Population Biology of Infectious Diseases. Dahlem Konferenzen. Springer-Verlag, Heidelberg. *To appear.*
HOEPRICH, P.D. 1977. Pp. 34-45. *In* Hoeprich, P.D. (ed.), Infectious Diseases. Harper and Row, New York.
HOGG, W.H., C.E. HOUNAM, A.K. MALLIK, and J.C. ZADOCS. 1969. Meteorological factors affecting the epidemiology of wheat rusts. Technical Note #99, World Meteorological Organization, Geneva.
JACOB, F. 1977. Evolution and tinkering. Science, 196:1161-1166.
JAENIKE, J. 1978. An hypothesis to account for the maintenance of sex within populations. Evolutionary Theory, 3:191-194.
JANZEN, D.H. 1980. When is it coevolution? Evolution, 34: 611-612.
JAYAKAR, S.D. 1970. A mathematical model for interaction of frequencies in a parasite and its host. Theoretical Population Biology, 1:140-164.
JENSEN, N.F. 1965. Multiline superiority in cereals. Crop Science, 5:566-568.
KARLIN, S., and J.L. MCGREGOR. 1972. Polymorphisms for genetic and ecological systems with weak coupling. Theoretical Population Biology, 3:210-238.
KEMPER, J.T. 1982. The evolutionary effect of endemic infectious disease: continuous models for an invariant pathogen. Journal of Mathematical Biology. *To appear.*
KIMURA, M. 1958. On the change of population fitness by natural selection. Heredity, 12:145-167.
KIMURA, M. 1965. A stochastic model concerning the maintenance of genetic variability in quantitative characters. Proceedings of the National Academy of Sciences, USA, 54: 731-736.
KIMURA, M., and G.H. WEISS. 1964. The stepping stone model of population structure and the decrease in genetic correlation with distance. Genetics, 49:561-576.
LEONARD, K.J. 1977. Selection pressures and plant pathogens. Annals of the New York Academy of Sciences, 281:207-222.
LEONARD, K.J., and R.J. CZOCHOR. 1978. In response to "Selection pressures and plant pathogens: stability of equilibria." Phytopathology, 68:971-973.
LEONARD, K.J., and R.J. CZOCHOR. 1980. Theory of genetic interactions among populations of plants and their pathogens. Annual Review of Phytopathology, 18:237-258.
LEVIN, B.R., A.C. ALLISON, H.J. BREMERMANN, L.L. CAVALLI-SFORZA, B.C. CLARKE, R. FRENTZEL-BEYME, W.D. HAMILTON, S.A. LEVIN, R.M. MAY, and H.R. THIEME. 1982. Evolution in host-parasite systems. *In* Anderson, R.M., and R.M. May (eds.),

Population Biology of Infectious Diseases. Dahlem Konferenzen. Springer-Verlag, Heidelberg. *To appear.*

LEVIN, S.A. 1972. A mathematical analysis of the genetic feedback mechanism. American Naturalist, 106:145-164. (Erratum, 1973, 104:320).

LEVIN, S.A. 1974. Dispersion and population interactions. American Naturalist, 108:207-228.

LEVIN, S.A. 1976. Population dynamic models in heterogeneous environments. Annual Review of Ecology and Systematics, 7:287-310.

LEVIN, S.A. 1978. On the evolution of ecological parameters; pp. 3-26. *In* Brussard, P.F. (ed.), Ecological Genetics: The Interface. Springer-Verlag, Heidelberg.

LEVIN, S.A. 1981. Mechanisms for the generation and maintenance of diversity in ecological communities; pp. 173-194. *In* Hiorns, R.W., and D. Cooke (eds.), The Mathematical Theory of the Dynamics of Biological Populations II. Oxford Symposium on the Dynamics of Populations. Academic Press, London.

LEVIN, S.A. 1983. Coevolution. *In* Freedman, H. (ed.), Population Biology. *To appear.*

LEVIN, S.A., and D. PIMENTEL. 1981. Selection of intermediate rates of increase in parasite-host systems. American Naturalist, 117:308-315.

LEVIN, S.A., and L.A. SEGEL. 1982. Models of the influence of predation on aspect diversity in prey populations. Journal of Mathematical Biology, 14:253-284.

LEVIN, S.A., and J.D. UDOVIC. 1977. A mathematical model of coevolving populations. American Naturalist, 111: 657-675.

LEWIN, R. 1982. Biology is not postage stamp collecting. Science, 216:718-720.

LEWIS, J.W. 1978. Maintenance of genetic polymorphism for two species in a host pathogen relationship. Masters Thesis, Iowa State University, Ames.

LEWIS, J.W. 1981a. On the coevolution of pathogen and host: I, General theory of discrete time coevolution. Journal of Theoretical Biology, 93:927-952.

LEWIS, J.W. 1981b. On the coevolution of pathogen and host: II, Selfing hosts and haploid pathogens. Journal of Theoretical Biology, 93:953-985.

LEWONTIN, R.C. 1970. The units of selection. Annual Review of Ecology and Systematics, 1:1-18.

LEWONTIN, R.C. 1977. Adaptation. Enciclopedia Einaudi Turin, 1:198-214.

LEWONTIN, R.C. 1980. Models of natural selection. *In* Barigozzi, C. (ed.), Vito Volterra Symposium on Mathematical Models in Biology. Lecture Notes In Biomathematics 38.

Springer-Verlag, Heidelberg.

LONGINI, I.M., JR. 1982. Model of epidemics and endemicity in genetically variable host populations. *Manuscript.*

MALÉCOT, G. 1948. Les Mathématiques de l'Hérédité. Masson et Cie, Paris. (English translation, 1969, W.H. Freeman, San Francisco).

MARUYAMA, T. 1971. The rate of decrease of heterozygosity in a population occupying a circular or linear habitat. Genetics, 67:437-454.

MAY, R.M., and R.M. ANDERSON. 1979. Population biology of infectious diseases: II. Nature, 280:455-461.

MAYNARD SMITH, J. 1976. Evolution and the theory of games. American Scientist, 64:41-45.

MAYNARD SMITH, J. 1977. Evolution and theory of games. *In* Matthews, W. (ed.), Mathematics in the Life Sciences. Lecture Notes in Biomathematics. Springer-Verlag, Berlin-New York.

MAYR, E. 1982. The Growth of Biological Thought. Diversity, Evolution, and Inheritance. Harvard University Press, Cambridge, Massachusetts.

MODE, C.J. 1958. A mathematical model for the coevolution of obligate parasites and their hosts. Evolution, 12:158-165.

MODE, C.J. 1960. A model of a host-pathogen system with particular reference to the rusts of cereals; pp. 84-96. *In* Biometrical Genetics. Pergamon Press, New York.

MODE, C.J. 1961. A generalized model of a host-pathogen system. Biometrics, 17:386-404.

MYERS, K., I.D. MARSHALL, and F. FENNER. 1954. Studies in epidemiology of infectious myxomatosis of rabbits. III. Observations on two succeeding epizootics in Australian wild rabbits on the Riverine Plain of south-eastern Australia 1951-1953. Journal of Hygiene, 52(3):337-360.

NAGYLAKI, T. 1977. Selection in one- and two-locus systems. Lecture Notes in Biomathematics. Springer-Verlag, Heidelberg.

NAGYLAKI, T. 1978. The geographical structure of populations, pp. 588-623. *In* Levin, S.A. (ed.), Studies in Mathematical Biology II: Populations and Communities. Mathematical Association of America, Washington, D.C.

PERSON, C. 1959. Gene-for-gene relationships in host-parasite systems. Canadian Journal of Botany, 37:1101-1130.

PERSON, C. 1966. Genetic polymorphism in parasitic systems. Nature, 212:266-267.

PIMENTEL, D. 1961. Animal population regulation by the genetic feedback mechanism. American Naturalist, 95:65-79.

PIMENTEL, D. 1968. Population regulation and genetic feedback. Science, 159:1432-1437.

PIMENTEL, D. 1982. Genetic diversity and stability in parasite-host systems. *Manuscript.*

RAND, A.S. 1967. Predator-prey interactions and the evolution of aspect diversity. Atas do Simposio Sobra a Biota Amazonica, 5(Zoologia):73-83.
RICKLEFS, R., and K. O'ROURKE. 1975. Aspect diversity in moths: a temperate-tropical comparison. Evolution, 29: 313-324.
ROCKLIN, S.M., and G.F. OSTER. 1976. Competition between phenotypes. Journal of Mathematical Biology, 3:225-261.
ROSS, J. 1982. Myxomatosis: the natural evolution of the disease. *In* Animal Disease in Relation to Animal Conservation. Symposium of the Zoological Society of London. In press.
ROUGHGARDEN, J. 1972. Evolution of niche width. American Naturalist, 106:683-718.
SAUNDERS, I.W. 1980. A model of myxomatosis. Mathematical Biosciences, 48:1-16.
SEDCOLE, J.R. 1978. Selection pressures and plant pathogens: stability of equilibria. Phytopathology, 68:967-970.
SIDHU, G.S. 1975. Gene-for-gene relationships in plant parasitic systems. Scientific Progress, Oxford, 62:467-483.
SLATKIN, M. 1970. Selection and polygenic characters. Proceedings of the National Academy of Sciences, USA, 66: 87-93.
STAKMAN, E.C., and C.M. CHRISTENSEN. 1946. Aerobiology in relation to plant disease. Botanical Review, 12(4):205-253.
VAN DER PLANK, J.E. 1975. Principles of Plant Infection. Academic Press, New York.
WADE, M.J. 1978. A critical review of the models of group selection. Quarterly Review of Biology, 53:101-114.
WALLACE, B. 1968. Topics in Population Genetics. W.W. Norton, New York.
WILSON, D.S. 1977. Structured demes and the evolution of group-advantageous traits. American Naturalist, 111:157-185.
WRIGHT, S. 1949. Adaptation and selection; pp. 365-389. *In* Jepson, G.L., G.G. Simpson, and E. Mayr (eds.), Genetics, Paleontology, and Evolution. Princeton University Press, Princeton.
WRIGHT, S. 1955. Classification of the factors of evolution. Cold Spring Harbor Symposia on Quantitative Biology, 20: 16-24.
YU, P. 1972. Some host-parasite genetic interaction models. Theoretical Population Biology, 3:347-357.

LIMITS TO SPECIALIZATION AND COEVOLUTION IN PLANT-ANIMAL MUTUALISMS

Douglas W. Schemske

Department of Biology,
and Committee on Evolutionary Biology
University of Chicago
Chicago, Illinois

Overemphasis on the role of coevolution in mutualisms has deflected attention from the evolutionary consequences of mutualistic interactions. In this paper, examples from a number of plant-animal mutualisms, including ant-guard, pollination and seed dispersal systems illustrate ecological and genetic constraints on 1) the potential for coevolution, and 2) the evolution of specialized mutualisms. Ant-guard systems, in which plants supply food, nest sites, or both, for ants in "exchange" for herbivore/predator control, vary widely in specificity. Plant species that provide "major" resources, i.e., nest sites and diverse foods that are predictable in time and space, generally are hosts to one or several ant species. Ant-plant specialization in these cases is generally extreme, and probably the result of coevolution. Where the plant provides a more limited resource, e.g., extrafloral nectar, the mutualism is very generalized. Selection for mutualism specificity in ant-plant systems is largely dependent on the extent to which the plant provides essential resources.

In plant-pollinator interactions, the selective effects of a mutualism are often highly asymmetrical where specificity may not reflect coevolution, as appears to be the case for the diverse and highly specialized bee fauna of North America. In general, the influence of a pollinator on plant fitness includes 1) a fecundity component, which is the efficiency of ovule fertilization, and 2) a "genetic" component which represents the variation in fitness among progeny attributable to pollen source. Individual and/or population variation in characters such as flower number, flower morphology, and nectar output may influence the composition of the pollinator fauna, and the distance of pollen flow. This approach to the evolution of pollination systems emphasizes the role of plant population structure, and the consequences of pollinator movements within and between populations.

Analogous to pollination systems, the effectiveness of seed dispersal by animals includes an "efficiency" component, determined by the percent of fruit removed, and a "quality" component, which is the proportion of dispersed seeds located in a site favorable to germination. Limitations to the specificity of seed-dispersal mutualisms include ecological constraints on seed size, temporal and spatial variation in seed germination sites, and the relatively broad level of habitat selection by most dispersal agents as compared to the microspatial distribution of germination sites.

Ecological and genetic approaches to the study of mutualisms, and their evolution, are emphasized. In particular, future studies must quantify 1) the extent of genetic variation in characters involved in a mutualistic interaction, 2) the ecological and genetic constraints on selection intensities, and 3) the proportion of the total variation in fitness which is attributable to the mutualism.

Coevolution is one of the most important processes influencing the patterns of variation and adaptation in the world's fauna and flora. It is most unfortunate, therefore, that the term coevolution is often misused and ill-defined. As Janzen (1980) has pointed out, coevolution has been incorrectly synonymized with "interactions," "mutualism," "plant-animal interaction," and other processes which may or may not reflect coevolution. This has probably arisen because the mechanisms of coevolution were prematurely invoked to explain a broad range of extremely complex biological phenomena. Such confusion distorts our abilities to formulate and test hypotheses or provide synthetic treatments of the topic, and comes when the literature on coevolution is greatly expanding (fig. 1), indicating the need for a rigorous use of the term.

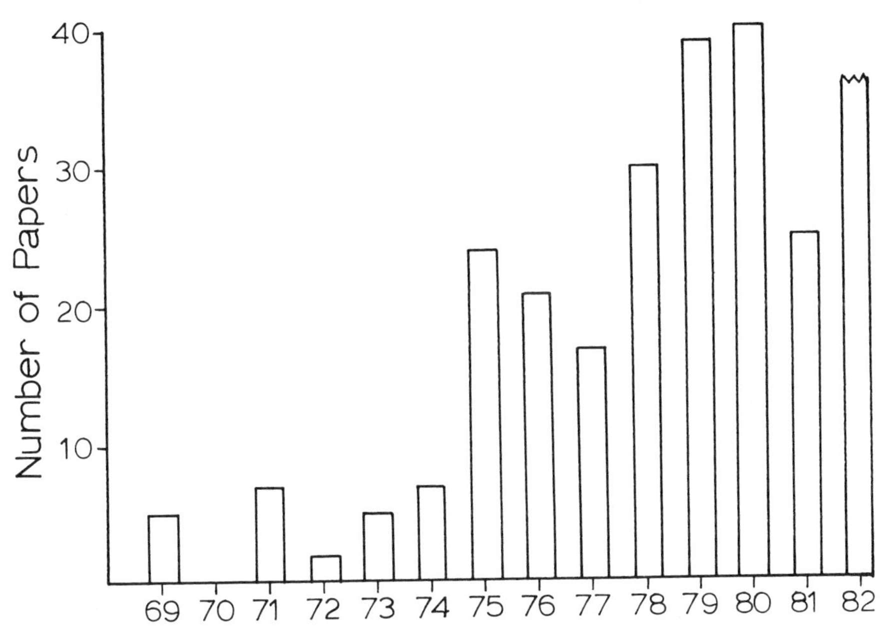

FIGURE 1. Number of published papers with "coevolution" or "co-evolution," as a key word used for indexing in Biological Abstracts.

A fallacy continually perpetuated is the notion that mutualisms are coevolved. A mutualism is generally regarded as any mutually advantageous association or interaction between dissimilar taxa. Although mutualisms provide significant positive effects on fitness for the taxa involved, this does not imply that coevolution has occurred. In this paper I outline some of the major questions in the area of plant-animal mutualisms, with particular reference to the role of coevolution in the structure, function and evolution of mutualisms. These questions are addressed utilizing examples from a number of plant-animal mutualisms, including ant-plant, pollination and seed dispersal systems. First, I will discuss various definitions of coevolution, then specifically examine the dynamics and evolution of mutualisms.

COEVOLUTION: A PROBLEM OF DEFINITION

In perhaps the most widely-read textbook in ecology, Ricklefs (1979) defined coevolution as the "development of genetically determined traits in two species to facilitate some interaction, usually mutually beneficial." This definition has several drawbacks. First, many examples of coevolution involve negative interactions between taxa, e.g., predator versus prey. Second, coevolution results *from* the interaction, not to facilitate it. Janzen (1980) provided the following definition: "Coevolution may be usefully defined as an evolutionary change in a trait of the individuals in one population in response to a trait of the individuals of a second population, followed by an evolutionary response by the second population to the changes in the first." This places unnecessary importance on the notion of reciprocity; coevolution can operate simultaneously or sequentially among interacting taxa. Roughgarden (1979) states that coevolution is "the simultaneous evolution of interacting populations." This definition is clear, and operationally correct, but perhaps does not sufficiently emphasize the point made by both Ricklefs and Janzen that coevolution results from interactions between taxa at the level of particular traits. Kiester, Lande, and Schemske (in preparation) define coevolution as "the joint selective effects on characters of interacting taxa, based on heritable variation in these characters." This definition includes positive and negative effects of an interaction and emphasizes that coevolution is dependent upon genetic variation in characters relevant to the interaction, and those characters which are genetically correlated with the selected characters.

Kiester et al. (op. cit.) also distinguish two different modes of coevolution, "uncoupled character coevolution" and

I. UNCOUPLED CHARACTER COEVOLUTION

JOINT SELECTIVE EFFECTS ARE PRESENT AT THE LEVEL OF INDIVIDUALS, BUT NOT AT THE LEVEL OF CHARACTERS.

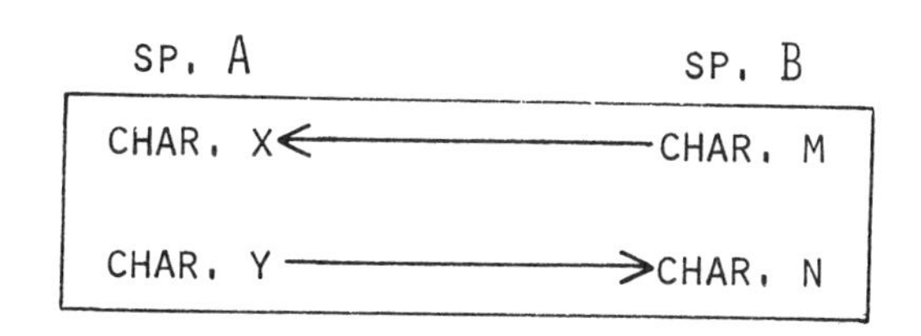

II. COUPLED CHARACTER COEVOLUTION

SELECTIVE EFFECTS BETWEEN TAXA ARE PRESENT AT THE LEVEL OF INDIVIDUALS AND CHARACTERS.

FIGURE 2. Modes of coevolution (from Kiester, Lande and Schemske, in prep.). Arrows in the examples indicate the direction of selection. See text for explanation.

"coupled character coevolution" (fig. 2). In the former, joint selective effects are present at the level of individuals, but not at the level of characters. This results from independent, unidirectional interactions between characters of interacting taxa. The absence of a response to selection in one species from a given interaction may result from lack of genetic variation for that character, or simply, from an absence of reciprocal selective forces. In coupled character coevolution, selective effects between taxa exist at the level of characters, and individuals (fig. 2). This mode of coevolution is probably the most obvious, in that the interaction at the level of characters directly influences fitness components of the interacting taxa. Note that in the uncoupled case, the presence of coevolution would probably be very difficult to distinguish empirically from directional selection, unless many characters, and their interactions, were monitored. In addition, where genetic correlations exist among characters in a given taxon, uncoupled character coevolution becomes a special case of coupled character coevolution (Kiester et al., in prep.).

The motivation for differentiating between these two modes was simply to describe the ways coevolutionary forces may act. The response to selection depends on the amount of genetic variability in selected characters, a critical component for which there is virtually no information, despite the enormous interest in coevolutionary processes. Focusing attention on interactions at the level of characters provides a quantitative approach to assessing the evolution of interacting taxa (see Roughgarden, 1979; Kiester et al., in prep.).

MUTUALISMS

May (1976) stated that of all interspecies interactions, mutualisms are perhaps the least understood, and that we lack the empirical and theoretical tools with which to develop a synthesis of mutualistic interactions. The many approaches used in the study of mutualisms indicate the complexity of the topic. Theoretical approaches have generally emphasized the demographic consequences of mutualistic interactions, on population, species or community levels (May, 1973, 1976; Levin and Udovic, 1977; Vandermeer and Boucher, 1978; Goh, 1979; Heithaus et al., 1980; Addicott, 1981), while the genetic consequences of mutualistic interactions have received less attention (but see Roughgarden, 1979; Kiester et al., in prep.). Empirical studies range from purely descriptive, natural-history accounts to experiments which quantify the benefits of mutualisms (Culver and Beattie, 1978; Schemske, 1980a; Horvitz, 1981).

Future research should integrate these various approaches and answer the following questions:

1. Is there genetic variation for traits that influence interactions among mutualists?

2. What proportion of the total variation in fitness among individuals in a given mutualist taxon is attributable to mutualistic interactions?

3. How are the positive effects of a mutualism associated with particular fitness components?

4. What ecological characteristics influence the evolution of mutualism structure and specificity?

5. How important is coevolution in the evolution of mutualistic interactions?

Unfortunately, we have no information relating to question 1; likewise, there are few data pertaining to question 2. For many mutualistic systems we can estimate single-character fitness components, as posed in question 3, but we have no integrated analysis of mutualism contribution to total fitness. The most significant advances in the area of mutualistic interactions will probably arise from studies which combine ecological and genetic approaches, and important contributions may come in the next decade (see May, 1982).

Questions 3 through 5 above focus on the ecological and evolutionary processes that influence mutualism structure, function and specificity. In plant-animal interactions, specificity is often incorrectly assumed to indicate coevolution and reciprocal selective forces. Where mutualistic associations are present, the potential for specialization and perhaps coevolution, may be limited by 1) the intensity of selection, 2) the extent of genetic variation in selected characters, and 3) ecological constraints that reduce selection intensities. I will now discuss how the fitness consequences of mutualistic interactions and the ecological characteristics of mutualists may influence mutualism evolution, and the potential for specificity and coevolution.

Ant-Plant Mutualisms

Many observational and experimental studies have verified the mutualistic components of ant-plant interactions (Janzen, 1966, 1969, 1974; Hocking, 1970; Bentley, 1977a and b; Tilman, 1978; Schemske, 1980a; Keeler, 1981). The plant provides an array of resources, including foods and nest sites, and ants, through their foraging and defense of space, provide defense against herbivores or fecal deposits that stimulate plant growth. Ant-plant mutualisms can be categorized in two classes, based upon the nature of the plant resource: 1) the extrafloral nectary (EFN) system, in which plants secrete a sugar solution, generally rich in amino acids, or 2) a "major resource" system, where plants provide a) foods rich in lipids or proteins, or b) domatia (cavities suitable for nesting), or both, occasionally in addition to EFN.

To what extent do the resource-based differences observed between these two classes of ant-plant mutualisms influence the specificity of the interaction? Table 1 summarizes data on the ant taxa observed at plants which provide limited resources, primarily EFN, and table 2 presents analogous information for plants with "major" resources. The difference in the apparent levels of ant-plant specificity is striking, with a range of four to 12 ant genera and five to 24 ant species observed on plants in EFN-based systems, as compared

Table 1. Ant assemblages at plants with limited resources, primarily extrafloral nectar (EFN).

Plant	Plant Resources			Ants		Locality	Source
	EFN	"Other" Foods[a]	Domatia	No. Genera	No. Species		
Helianthella quinquenervis	+	-	-	3	5	Colorado	Inouye and Taylor 1979
Ipomoea pandurata	+	-	-	7	10	N Carolina	Beckman and Stucky 1981
Pteridium aquilinum	+	-	-	7	8	New Jersey	Tempel, pers. comm.
Mentzelia nuda	+	-	-	4	5	Nebraska	Keeler 1981
Costus woodsonii	+	-	-	4	7	Panama	Schemske 1980a
C. pulverulentus	+	-	-	9	12	Panama	Schemske 1982
C. scaber	+	-	-	9	13	Panama	Schemske 1982
C. allenii	+	-	-	11	17	Panama	Schemske 1982
C. laevis	+	-	-	12	24	Panama	Schemske 1982
Ipomoea carnea	+	-	-	6	10	Costa Rica	Keeler 1977
Calathea ovandensis	+	-	-	9	16	Mexico	Horvitz, pers. comm.
Bixa orellana	+	-	-	8	12	Costa Rica	Bentley 1977a
Aphelandra deppeana	+	-	-	ND[b]	8	Costa Rica	Deuth 1977
Ferocactus gracilis	+	-	-	10	11	Mexico	Blom and Clark 1980
Ochroma pyramidale	+	+[c]	-	4	ND[b]	Costa Rica	O'Dowd 1979

[a] Protein and/or lipid-rich, solid foods
[b] No data
[c] Pearl bodies

Table 2. *Ant assemblages at plants with "major" resources.*

Plant	Plant Resources			Ants		Locality	Source
	EFN	"Other" Foods[a]	Domatia	No. Genera	No. Species		
Codonanthe crassifolia	+	+	-[b]	1	1	Costa Rica	Kleinfeldt 1978
Acacia cornigera	+	+	+	1	1	Mexico	Janzen 1967
A. drepanolobium	+	+	+	1	3	East Africa	Hocking 1970
Cecropia spp.	-	+	+	1	5	Central America	Janzen 1969
Macaranga spp.	-	+	+	1	1	Southeast Asia	Rickson 1980
Solanopterus brunei	-	+	+	4[c]	6[c]	Costa Rica	Gomez 1974
Piper cenocladum	-	+	+				
P. fimbriulatum	-	+	+	1	1	Costa Rica	Risch et al. 1977
P. sagittifolium	-	+	+				
Hydnophytum formicarium	-	-	+				
Myrmecodia tuberosa	-	-	+	1	1	Sarawak, Malaysia	Janzen 1974
Phymatodes sinuosa	-	-	+				
Dischidida rafflesiana	-	-	+				
Barteria fistulosa	-	-	+	1	2	West Africa	Janzen 1972a
Triplaris cumingiana	-	-	+	1	1	Panama	Croat 1978
Tillandsia butzii	-	-	+	7	ND[d]	South America	Benzing 1970
T. caput-medusae	-	-	+	3	ND[d]	South America	Benzing 1970

[a] Protein and/or lipid-rich, solid foods

[b] Plant roots may support nests

[c] Gomez (1974) reported that a species of Azteca was probably restricted to this plant, but that other ants were also observed.

generally to a single ant genus and one to several ant species at plants with major resources. Although major resource mutualisms (table 2) may often involve ants that visit more than one plant species (Huxley, 1980), these ant-plant mutualisms are certainly more restricted than those involving plants which only produce EFN (table 1).

EFN supplies only a small fraction of the nutritional requirements of an ant colony, and limits the selection intensity on ant characters that may increase ant-plant specificity. Where plants provide complete resources, we expect greater potential for ant-plant specificity, and coevolution. It appears that the extent of ant-plant specificity is directly related to the predictability and complexity of resources. This is a striking example of the interaction between ecological components of mutualism structure, and the potential for specialization and coevolution. Where complete resources are provided by plants, the fitness consequences to ant colonists of further specialization may provide the selection intensity required to promote coevolution among mutualists.

This poses an important question that has not previously been addressed in mutualism studies, i.e., to what extent are mutualists coevolved? In the ant-plant system, ant species may vary in their protective benefits to the plant, and therefore, plant characters may be under selection to attract and maintain particular ant taxa. Coevolution could occur if there was genetic variation for plant resources and ant preferences, and if the selective effects of the interactions were mutually beneficial. Consistent preferences by ants and positive selective effects among mutualists could lead to increased specialization.

To investigate the potential for ant-plant specialization in EFN-based systems, I quantified the composition of ant assemblages at the EFNs of four neotropical *Costus* spp. in Panama. The objectives were to determine the among-plant species variation in ant assemblages, and the ecological correlates of ant assemblages. With respect to the former, there were striking differences among *Costus* species in ant-species richness, with 12 ant species on *Costus pulverulentus*, 13 on *C. scaber*, 17 on *C. allenii* and 24 on *C. laevis* (Schemske, 1982). Thus within a single plant genus, there is significant variation in the structure and specificity of the ant-plant mutualism.

To identify the major characteristics of the interaction that influence ant assemblages, it was necessary to partition this among-plant species variation. There was no significant effect of nectar volume or composition, or plant density, on ant species richness. However, there was a highly significant effect of the vertical distribution of inflorescences among plant species, and its interaction with ant foraging ranges. A positive correlation was observed between the number of ant species on a given *Costus* and inflorescences height, all computed on plant species means. More importantly, the among-plant species differences are a function of ant habitat preferences. Terrestrial ant species are virtually equal in abundance for all *Costus* spp., while arboreal ants increase in frequency with inflorescence height (Schemske, 1982).

This suggests that ant occupation of EFNs in this system is largely a stochastic process, based on the interaction between inflorescence height and ant foraging ranges. It is unlikely that the proportion of the total selection intensity on inflorescence height due to the ant-plant mutualism has been sufficient to influence the evolution of this plant character. Selection on inflorescence height includes components due to pollination, seed dispersal and other ecological and physiological factors. There is no evidence that coevolution has been important in the dynamics of this mutualistic interaction. The potential for ant-plant specialization, and perhaps coevolution, is probably limited by the mosaic distribution of ant territories (see Leston, 1978), the spatial and temporal unpredictability of EFN production, and the limited resources provided by plants with EFNs.

Pollination Systems

Mutualisms and coevolution have been important to varying degrees in the evolution of pollination systems. Analogous to the discussion of ant-plant systems, an important question is, what constitutes plant-pollinator coevolution? The diversity and host-plant specialization among North American bees will serve as an example. There are 3,465 species of bees (Apoidea) in America north of Mexico (Krombein et al., 1979), but, excluding gymnosperms and grasses, only 14,575 species of flowering plants (J.T. Kartesz, pers. comm.), a ratio of plant to bee species of only 4.2. This apparent pollinator richness may conjure up visions of intense plant-pollinator coevolution, with the speciation of pollinators and hosts dependent upon selection pressures resulting from the interaction. My purpose, as indicated in the introduction, is to emphasize that "interaction" is not synonymous with coevolution, and that plant-animal interactions can be highly asymmetrical in their selective effects, as appears to be the case for this system.

With respect to pollen preferences, bees can be classified as polylectic or oligolectic. Polylectic species collect pollen from a range of unrelated plants while oligolectic species consistently collect pollen from a single species, or a group of related species (Linsley, 1958). Oligolecty often involves physiological and morphological adaptations to particular plant taxa (Linsley, 1958; Thorp, 1969; Cruden, 1972). As emphasized by others (Michener, 1954; Heithaus, 1979), the term "oligolecty" does not adequately express the full range of bee foraging. For example, many bees are locally or temporally specialized to particular pollen sources (Stephen et al., 1969). Floral constancy, i.e., specialization by individual bees in a polylectic species (Grant, 1950; Linsley, 1958; Heinrich, 1976), is very common, particularly among social species. Although the "best" data for comparing levels of specialization among bee species would include temporal flexibility, and within and between population differences in flower visitation, this information is very difficult to obtain, and therefore very scarce. In contrast, data on pollen sources have been summarized for approximately one-quarter of all bee species in North America (Krombein et al., 1979), providing a rich source of comparative material.

In this section I ask: 1) What is the frequency and intensity of specialization in North American bees? 2) Do bee taxa vary in their levels of specialization, and 3) Is host-plant specificity indicative of coevolution between plants and pollinators? Data pertaining to questions one and two were summarized from the comprehensive list of flower visitation, pollen sources and geographic distributions collated by Dr. Paul Hurd for North American bees (Krombein et al., 1979). Only bee species indicated as polylectic, or oligolectic at the level of plant family, genus, or species were included in this analysis.

A summary of pollen preference for North American bees indicates high levels of host-plant specialization; two-thirds (64%) of the 960 bee species for which adequate data exist are oligolectic at some level (table 3). Although these data are

Table 3. Taxonomic distribution of pollen preferences in North American bees. Oligolectic species are classified at the levels of plant family, genus or species. Numbers in parentheses give percent. Data summarized from Krombein et al. (1979).

Family	Polylectic	Oligolectic			Total
		Family	Genus	Species	
Adrenidae	69 (17.0)	39 (9.6)	209 (51.5)	89 (21.9)	406
Colletidae	10 (31.3)	11 (34.4)	8 (25.0)	3 (9.4)	32
Oxaeidae	1 (25.0)	0	0	3 (75.0)	4
Halictidae	21 (38.9)	2 (3.7)	23 (42.6)	8 (14.8)	54
Melittidae	1 (6.7)	2 (13.3)	8 (53.3)	4 (26.7)	15
Megachilidae	117 (58.5)	14 (7.0)	49 (24.5)	20 (10.0)	200
Anthophoridae	79 (39.1)	49 (24.3)	54 (26.7)	20 (9.9)	202
Apidae	47 (100.0)	0	0	0	47
Total:	345 (35.9)	117 (12.2)	351 (36.6)	147 (15.3)	960

only an estimate, given the large number of bee species that are poorly known, they nevertheless illustrate the general phenomenon of considerable host-plant specialization in North American bees. For 14 different geographic regions in North America, Moldenke (1979a) estimated that the relative frequency of specialized bees (taxonomic level of host-plant specialization not given) averaged 32%, and ranged from 16% in tundra and muskeg to 67% in deserts. Specialization is more frequent in California, with 40 to 55% specialization in most regions (Moldenke, 1976).

A statistical analysis of pollen preference for the three best-known bee families with heterogenous host preferences, the Andrenidae, Megachilidae and Anthophoridae, indicates a highly significant difference among bee taxa (table 4). A direct comparison of the ratios of observed to expected number of bee species in a given pollen preference class indicates that the Andrenidae are far more specialized than the Megachilidae or Anthophoridae (table 5). The Andrenidae contain proportionately more species than expected by chance which are oligolectic at the level of plant genus or species, while the

Table 4. Distribution of pollen preferences for three bee families in North America with polylectic and oligolectic taxa. The expected number of species is given in parentheses. $\chi^2 = 156.5$, $p < 0.0001$. *Data from Krombein et al. (1979).*

Family	*Polylectic*	*Oligolectic*		
		Family	*Genus*	*Species*
Andrenidae	*69 (133.2)*	*39 (51.3)*	*209 (156.8)*	*89 (64.8)*
Megachilidae	*117 (65.6)*	*14 (25.2)*	*49 (77.2)*	*20 (31.9)*
Anthophoridae	*79 (66.3)*	*49 (25.5)*	*54 (78.0)*	*20 (32.3)*

Table 5. Ratios of observed to expected number of bee species, by pollen preference, calculated from data in table 4.

Family	*Polylectic*	*Oligolectic*		
		Family	*Genus*	*Species*
Andrenidae	*.52*	*.76*	*1.33*	*1.37*
Megachilidae	*1.78*	*.56*	*.63*	*.63*
Anthophoridae	*1.19*	*1.92*	*.69*	*.62*

Megachilidae and Anthophoridae have more polylectic species, and the Anthophoridae have more family oligoleges than expected (table 5). Although these data refer only to pollen sources, the levels of specialization observed in pollen collection parallel those for total floral visitation, where either pollen or nectar are collected. A frequency distribution of number of plant genera visited for pollen or nectar (fig. 3) indicates that polylectic species and family oligoleges are much more generalized in their floral visitation than bee species oligolectic at the plant genus or species level.

Thus we can identify bee taxa that may exert markedly different selective forces on plant characters, based on pollen preferences (tables 3, 4 and 5) or total floral visitation (fig. 3). This may suggest the potential for plant-pollinator mutualism, specialization and coevolution. In fact, there is virtually no evidence that the tremendous diversification and specialization of North American bees is related to coevolution between plants and bees (Michener, 1979). Detailed studies of the plant family Onagraceae indicate that plant evolution has rarely been influenced by the foraging behavior of bees with restricted pollen preferences (Linsley et al.,

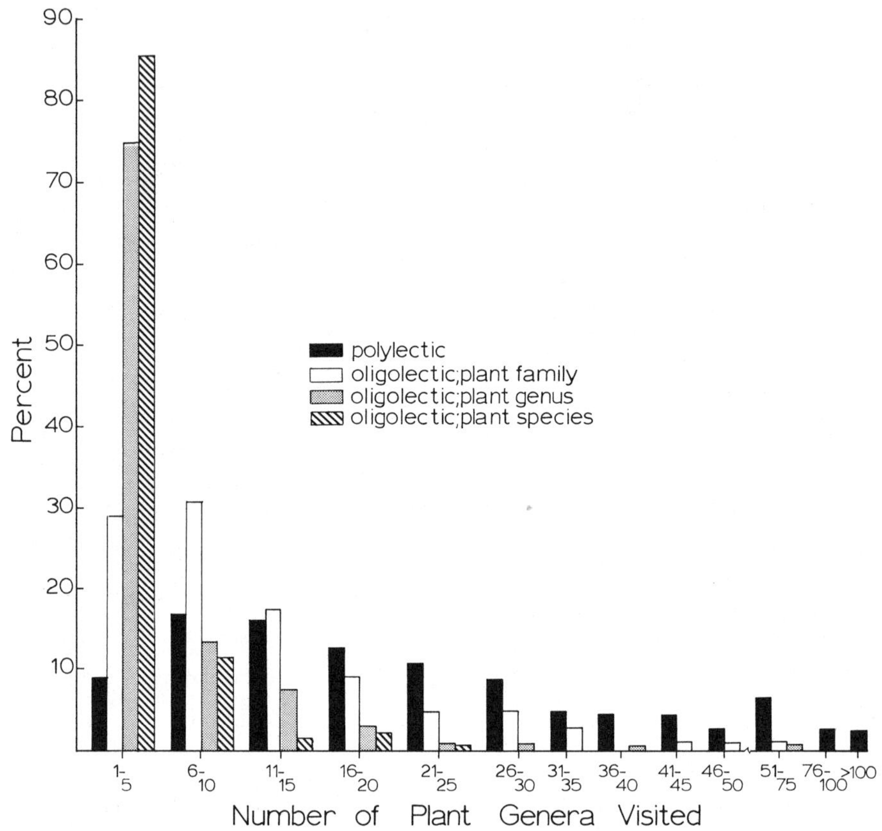

FIGURE 3. *Frequency distribution of number of plant genera visited for nectar or pollen by polylectic bees (solid bars), or bee species oligolectic at the level of plant family (open bars), genus (stippled bars) and species (cross-hatched bars). Except for the Apidae, which were excluded from the analysis, all data are based on the visitation records of bees included in table* 3, *summarized from Krombein et al. (1979).*

1973; MacSwain et al., 1973). For example, the plant genus *Camissonia* has over 20 species of bees restricted to either the genus or species level in pollen utilization, yet there is little evidence of plant-pollinator coevolution (Linsley et al., 1973). Moldenke (1979a) observed that many bee taxa which demonstrate host-plant specialization are not the primary pollinators of their hosts, and Michener (1979) provides several examples of specialized bees which play no role in pollination. Motten et al. (1981) found that the pollination efficiency of *Andrena erigeniae*, an oligolege of *Claytonia virginica*, was virtually identical to that of the fly *Bombylius major*, an extremely generalized forager.

For host-plant specialization by bees to have a significant effect on plant characters, we would expect some symmetry in frequency of interactions. The intensity of selection of bee foraging behavior on plant characters is reduced as the diversity and frequency of other pollinators increases. To what extent is bee specialization correlated with the specificity of plants to particular pollinators? In many cases, the high diversity of insect visitors to any particular plant species probably overrides the selective effects of particular plant specialists. This is certainly the case for bees specialized to the Compositate. Of the 117 bee species that are specialized to a given plant family (table 3), 108 are specific to the family Compositae. In addition, there are 40 species of bees specialized to Composite genera, and 13 bee species specialized to Composite species. Moldenke (1979a) estimates that more than 525 bee species are specialized at the level of Composite family or genus, and that this represents more than one-third of all specialized bees in North America. This impressive host-plant specialization is undoubtedly a function of the generalized inflorescence morphology of Composites, allowing easy access for pollen collection by many bee species. Despite the marked bee-to-plant specificity observed in the Compositae, individual plant species are extremely generalized with respect to insect visitors. Records of insect visitation to 55 species of composites, summarized from Robertson (1928), indicate modes of 11 to 15 bee genera, 11 to 20 bee species and 25 to 50 insect species per plant (fig. 4). Thus bee-to-plant specialization is common in composites, but plant-to-bee specialization is not. The conclusion is that the fitness consequences of bee-plant interactions may be highly asymmetrical, thus constraining the potential for plant-pollinator coevolution.

The diversification of North American bees, with a large number of pollen specialists, is a fascinating example of local evolution and host-plant switches (Moldenke, 1979a,b), but coevolution appears to have played a minor role. Linsley and MacSwain (1958) proposed two models to explain the processes of speciation in oligolectic bees. The "host-switch hypothesis" proposes that bee species with flexible pollen preferences may switch to different pollen sources during periods of pollen shortage. Where the pollen source is also the mating site, a host-plant switch may lead to reproductive isolation, and speciation. Similarly, Thorp (1969) suggests that host-plant switches within a population may lead to disruptive selection. The "fragmentation hypothesis" suggests that periodic local extinctions occur during years when the preferred pollen source is absent, leading to disjunct populations, and geographic speciation. A comparison of the geographic ranges of polylectic and oligolectic bees reveals a high percentage of specialized bees with restricted geographic distributions (fig. 5). The percent of bee species with ranges less than or equal to 15,000 sq. mi. is 2.0 for polylectic species, 1.8 for family-oligoleges, 14.2 for genus-oligoleges and 27.3 for species-oligoleges. Only 5.7% of all polylectic species have ranges less than or equal to 31,000

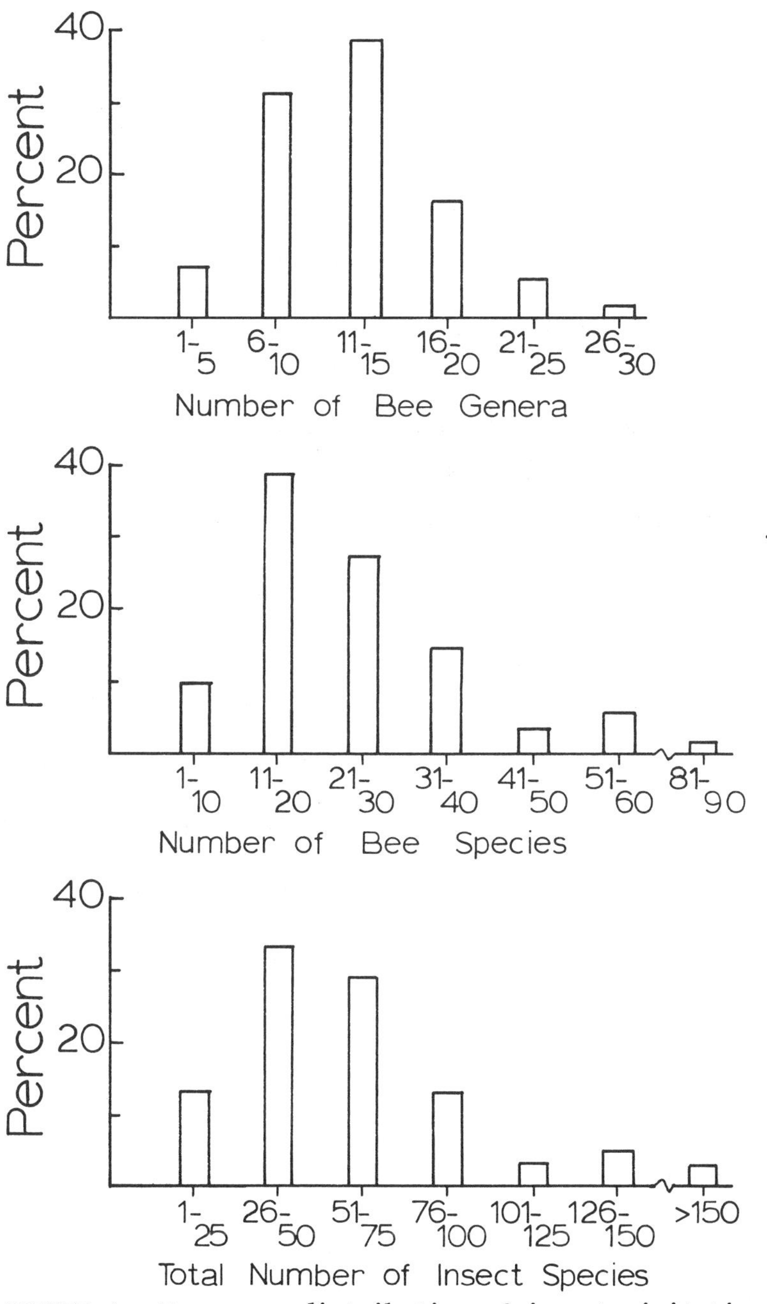

FIGURE 4. *Frequency distribution of insect visitation to 55 species of Compositae. Data are summarized from Robertson (1928), and are based on observations made within a 15 mi radius of Carlinville, Illinois.*

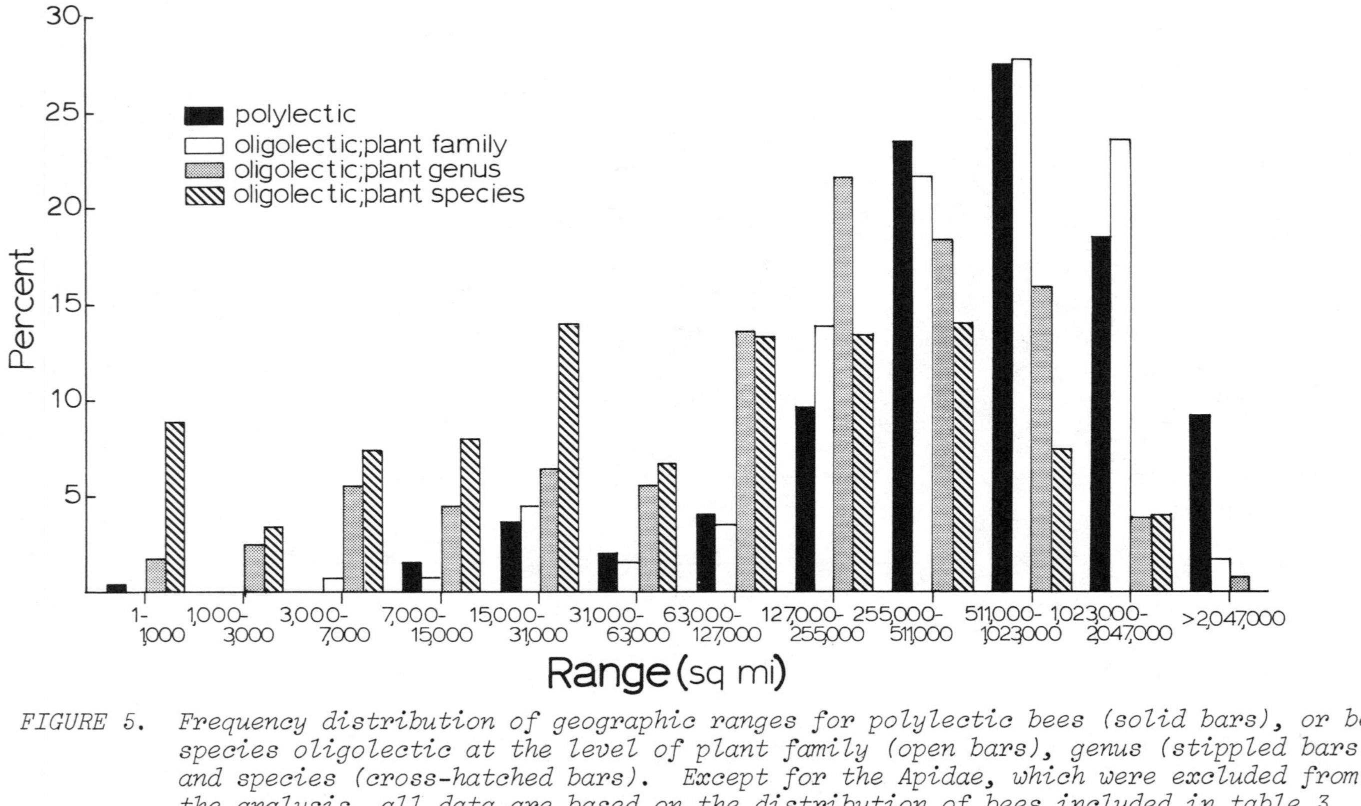

FIGURE 5. Frequency distribution of geographic ranges for polylectic bees (solid bars), or bee species oligolectic at the level of plant family (open bars), genus (stippled bars) and species (cross-hatched bars). Except for the Apidae, which were excluded from the analysis, all data are based on the distribution of bees included in table 3, summarized from Krombein et al., (1979).

sq. mi., as compared to 41.3% of all species oligoleges. Thus, host-plant specialization clearly limits geographic distribution and, as already indicated, may promote speciation by increasing the potential for geographic isolation.

This discussion of host-plant specialization in North American bees indicates the need for caution in studies of plant-pollinator interactions. Without a quantitative assessment of selection intensities on the characters involved in an interaction between taxa, coevolution cannot be assumed. At present, the data available for most plant-pollinator interactions are insufficient to distinguish between directional selection and coevolution as processes influencing the evolution of plant or pollinator characters. Specialization of floral characters may not indicate coevolution between plants and pollinators. For example, many violets (*Viola* spp.) lack specialized pollinators, despite the relatively complex floral morphology of the group (Beattie, 1974). The frequent observations that the morphologies of plant and pollinator "match", e.g., the correlation between beak length and corolla length in hummingbird-pollinated plants (Wolf et al., 1976) or the analogous association between tongue length and corolla length in bee-pollinated species (Inouye, 1980), are not sufficient evidence for coevolution (Macior, 1971), although they are often interpreted as such (Stiles, 1981). I suggest that much of the previous work which infers coevolution or coadaptation in pollination systems needs to be re-evaluated with greater attention given to a) the directions of selection, and b) the responses to selection in plant-pollinator interactions.

Fitness components in plant-pollinator interactions. These conclusions indicate, as suggested earlier, that we should treat interactions between taxa, in this case between plant and pollinator, as interactions between characters, and assess their net effects on fitness. The impact of a pollinator on plant fitness is then a function of the pollinator's effect on a suite of plant characters, and a regression of the plant characters on fitness (Kiester et al., in prep.). For plants we can ask, 1) What are the critical fitness components influenced by different pollination systems?, and 2) How might the interactions between characters of plants and pollinators promote coevolution?

With respect to question one, we can identify two important fitness components in plants. The most obvious is a fecundity component, here defined as the extent to which pollinators limit seed output. The second, and perhaps most important component of plant fitness, with respect to the evolution of pollination systems, concerns the genotypic composition and parentage of the progeny produced. Virtually all of the literature in pollination biology concerns the fecundity component of plant fitness. However, the extent to which fecundity is limited by pollinator services or resources is known for very few species. As reviewed by Bierzychudek (1981), seed production is pollinator-limited in some species (Schemske et al., 1978; Schemske, 1980a,b), and resource-limited in others (Willson and Price, 1977; Stephenson, 1979). Fecundity is probably influenced by both processes in many species (Schemske, 1977; Janzen et al., 1980; Willson and Schemske, 1980), but insufficient experimental data are available to test this claim (*contra* Stephenson, 1981). The genotypic component has never been directly explored in the context of plant *x* pollinator interactions.

The fundamental question is how selection on the fecundity and genotype components of plant fitness influences the evolution of pollination systems. Phenotypic variation in plant and floral characters will influence these components to varying degrees through effects on pollinator foraging and pollen movement, as summarized in figure 6. Pollinator visitation rate (fig. 6A) is influenced by: flowering time (Schemske, 1977; Zimmerman, 1980); the number of flowers per plant (Willson and Price, 1977; Schemske, 1980b and c); density and composition of flowering plants in a patch (Thomson, 1978); nectar quantity (Heinrich and Raven, 1972; Heinrich, 1975; Pyke, 1978); and by flower color (Levin, 1972; Mogford, 1974; Kay, 1976; Waser and Price, 1981). The number of pollen grains carried by individual pollinators is also expected to vary with plant phenotype (fig. 6B), but few data are available on this aspect of plant-pollinator interactions. Waddington (1981) observed that the number of pollen grains carried by bumblebees visiting *Delphinium* was greater after visits to nectar-rich inflorescences. The number of pollen grains deposited (fig. 6C) is influenced by nectar rewards, floral morphology and pollinator constancy. Thomson and Plowright (1980) found that the number of pollen grains deposited on flowers of *Diervilla lonicera* by bumblebees increased with nectar volume per flower. The effectiveness of pollination, i.e., the number of ovules fertilized (fig. 6D), depends upon the compatibility between pollen and seed parents (de Nettancourt, 1977), and the germinability of pollen grains. With respect to the latter, recent *in vitro* experiments on *Costus guanaiensis*, a neotropical herb, have demonstrated that an increase in the number of pollen grains per clump results in a significant increase in germination rate and pollen tube length (Schemske and Fenster, in press).

The plant and floral characters illustrated in figure 6 influence the number of seeds produced and the number, genotype and spatial distribution of pollen parents for a given seed crop. For example, variation in the numbers and positions of anthers and stigmas influences the number of pollen grains carried and deposited by pollinators, a fecundity

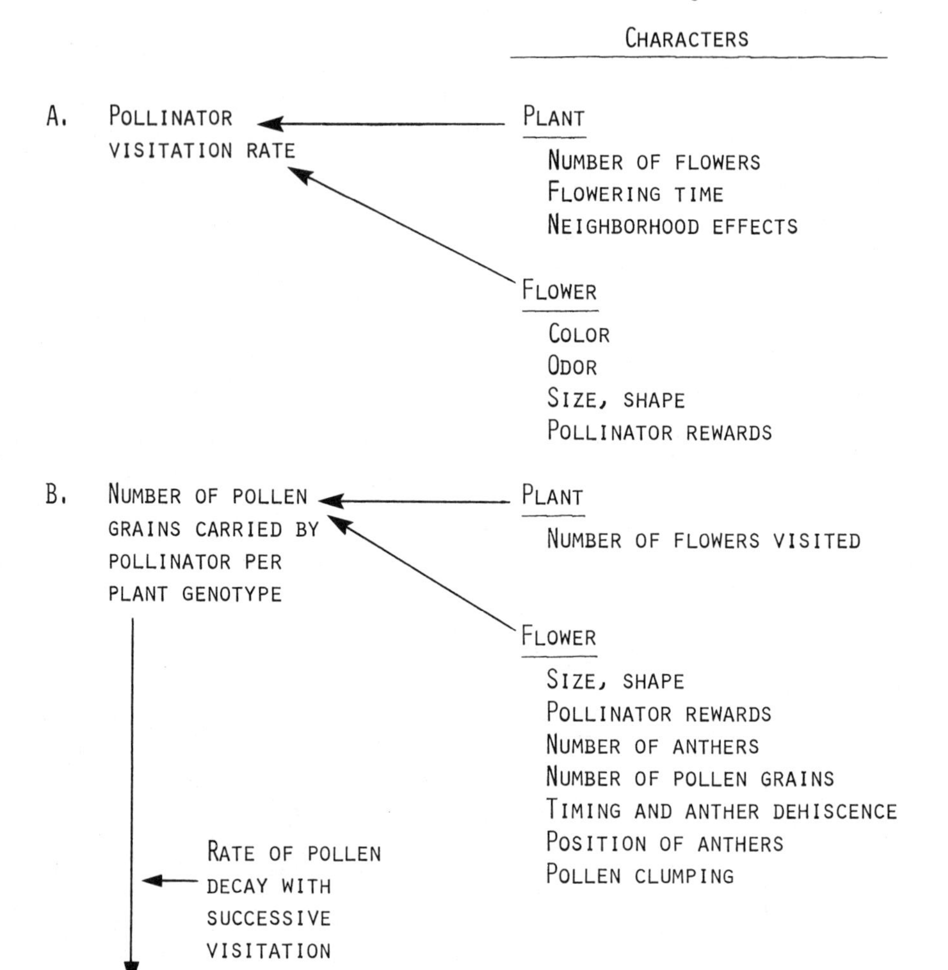

FIGURE 6. Effects of phenotypic variation in plant and floral characters on pollinator visitation and pollen movement.

component, and the diversity of pollen parents, a genotype component. The extent of pollen carryover has dramatic effects on the number of pollen parents that may fertilize the seed crop of a given female (Schemske, in prep.), influencing the genotypic diversity and fitness of progeny.

Pollinator taxa vary in their pollination effectiveness, thus floral morphologies may evolve to attract those pollinators which deposit an optimal number of pollen grains. The optimum is some function of 1) the probability of flower visitation, and 2) the importance of genetically variable progeny through parentage by multiple pollen donors. For *Campsis radicans*, Bertin (1982) observed that hummingbirds transferred 10 times more pollen grains per stigma per visit than honeybees or bumblebees. Similarly, honeybees are more efficient pollinators of *Oenothera fruticosa* than the beetle *Chauliognathus* (Primack and Silander, 1975).

An extremely important aspect of pollinator behavior which influences the genotype component of plant fitness is pollinator flight distance, and its effect on pollen flow. The evolutionary consequences of gene flow have been examined in two primary ways: 1) the effects of gene flow on effective population size (*Ne*), the number of individuals in a colony that mate at random (Wright, 1946), and 2) the fitness consequences of gene flow distance. For plants, the critical parameters affecting *Ne* are: 1) the variance in gene dispersal distance, including pollen and seed components, 2) flowering plant density, and 3) the proportion of seeds produced through outcrossing (Levin and Kerster, 1974). Estimates from plant populations often indicate that *Ne* is small, thus providing the potential for genetic differentiation of populations through drift (Kerster and Levin, 1968; Beattie and Culver, 1979; Schaal, 1980; Levin, 1981). An unfortunate consequence of the emphasis on estimating *Ne* is the assumption that the variance in gene dispersal distance is a constant for a given set of plant densities and outcrossing rates. With respect to pollination systems, pollinators vary in their flight distances and foraging behavior, thus providing the opportunity for selection on floral characters that could influence pollinator composition and pollen flow (fig. 6).

In this context, Schmitt (1980) analyzed the pollination of *Senecio* with respect to the flight distances of its two primary pollinators, bumblebees and butterflies. Most bumblebees flights were less than 1 m, while 95% of all flights by butterflies were more than 4 m. She concludes that if butterflies accounted for only 10% of all pollen dispersal, *Ne* would be an order of magnitude greater than that if bumblebees were the only pollinator. Similarly, Beattie and Culver (1979) find that the composition of pollinators visiting *Viola* significantly influences the distance of pollen flow, and *Ne*. These data clearly indicate that subtle variations in the composition of the pollinator fauna can have a major impact on *Ne*.

An important problem that has not been adequately addressed concerns the fitness consequences of pollen flow within and between populations. Where populations are genetically differentiated, pollen flow between populations may lead to greater progeny fitness due to heterosis if inbreeding is high within populations. This could result in selection on plant characters that influence pollinator composition and foraging behavior, and as a result, pollen flow distance. In this context, Zimmerman (1982) observed that bumblebees visiting plant species for pollen flew shorter distances than when foraging for nectar. Where characters that promote pollen flow exert a positive selective force on pollinator fitness, plant pollinator coevolution may occur.

Variation in the quality of floral rewards is more likely to have an impact on pollen flow distance via the composition of the pollinator fauna, than by changes in the foraging behavior and flight distance for particular pollinator species. For example, several studies have indicated that pollinator flight distance is inversely correlated with the quality of the reward in the flower previously visited (Pyke, 1978; Pleasants and Zimmerman, 1979; Waddington, 1980, 1981; Zimmerman, 1981). If this is generally true, an increase in reward quality (e.g., nectar volume and concentration) would decrease the mean flight distance of individual pollinators, but may attract other pollinator taxa with higher resource requirements (Heinrich and Raven, 1972), and flight distances.

There are few data available for determining the effects of pollinator flight distance on the fertility and genotype components of plant fitness. Price and Waser (1979) provide some evidence that distance of pollen flow influences seed production, a fecundity component, in *Delphinium nelsonii*. Seed set following hand pollination was greatest when pollen was obtained 1 to 100 m from the seed parent, as compared to shorter or longer distances. There was also a tendency for seedling survivorship, a genotype component of fitness, to be highest at intermediate outcrossing distances. They interpret these data as evidence for inbreeding, and outbreeding depression resulting from population differentiation on a microgeographic scale. Similarly, Levin (1981) reports that in *Phlox drummondii*, pollen germination and seed set increased, and seed abortion decreased, for interparent distances up to 10 m. Although the *Delphinium* and *Phlox* data strongly suggest that populations are genetically subdivided, with the genetic similarity between individuals correlated with their spatial relationships, the extent and scale of population differentiation are unknown.

Population structure, and the consequences of pollen flow on progeny fitness have recently been investigated in a series of greenhouse experiments with *Impatiens pallida*, a woodland annual (Schemske, in prep.). *Impatiens pallida* produces two different types of flowers: 1) cleistogamous (CL) flowers which never open and which are obligately self-pollinated, and 2) chasmogamous (CH) flowers which are large, nectar-rich and insect-pollinated (Schemske, 1978). Most populations of this species are selfing, with a high proportion of CL flowers. Thus the production of CH flowers cannot be interpreted as a response to selection for increased fecundity. To determine the fitness consequences of outcrossing in this system, I 1) quantified the extent of local, genetic differentiation in two populations, and 2) compared the fitness of selfed and outcrossed progeny, where the latter were derived from crosses within and between subpopulations. Population differentiation was examined by collecting 20 plants from each of three parallel transects (A,B + C) at two different localities, Brownfield and Allerton, in central Illinois. Transects were 32 m apart at Brownfield and 50 m apart at Allerton. CL seeds were collected from each plant, and families consisting of 10 plants each were grown in the greenhouse and scored for a number of quantitative characters, e.g., height, leaf length, flowering time, fecundity. A hierarchal analysis of variance with seed weight as a covariate indicated a highly significant effect of transect and seed parent for all characters at each locality. Thus, there was significant genetic variation within and among transects at each locality.

Population differentiation within the range of pollinator movement, as documented above, provides the background information required to test the hypothesis that interparent distances influence progeny fitness. To quantify the effects of pollen parentage on progeny fitness, four classes of seed were produced from plants of one transect (transect C) at both Brownfield and Allerton. These classes include 1) selfed seed from CL flowers (transect C, selfed), 2) outcrossed seed from within-transect pollinations (C x C), 3) outcrossed seed from pollinations with the adjacent transect (C x B), and 4) outcrossed seed from pollinations with the farthest transect (C x A). Plants from all classes were grown in the greenhouse in a completely randomized design, and total fecundity recorded. For both localities there was a significant effect of cross type, with selfed progeny producing the fewest pods (table 6). Outcrossing between transects resulted in higher fecundity than outcrossing within transects. The absolute and relative effect of interparent distance on fecundity was greatest for Brownfield, where outcrossing with the farthest transect resulted in a 59% increase in fecundity over that of selfed progeny, as compared to 13% for outcrossed progeny from within-transect pollinations (table 7). Further research is needed to determine if these fitness differences, based on greenhouse experiments, are similar in magnitude and direction to those in natural populations. Progeny produced from crosses between genetically-differentiated subpopulations may suffer outbreeding depression if local selection pressures are sufficiently strong to negate the positive effects of heterosis (Price and Waser, 1979).

In summary, the *Delphinium*, *Phlox*, and *Impatiens* studies indicate that a subdivided population structure could lead to specific selective pressures on pollination characteristics that influence pollen flow between subpopulations. Where progeny fitness is correlated with pollen flow distance we expect strong selection on floral characters that may attract specific pollinators. These characters could coevolve with pollinator preferences. By determining the fitness consequences

Table 6. *ANOVA and multiple comparison of means (Duncan's Multiple Range Test) for total pods per plant produced by* Impatiens pallida *derived from selfing, and outcrossing within and between populations. Main effects in the ANOVA are cross type (CROSS), seed parent (SDP), and the CROSS x SDP interaction. Seed weight (SDWT) is a covariate. "Pollen parents for outcrossed seed" refers to the source transects A, B and C at each locality, with plants from transect C serving as seed parents.*

						Multiple Comparisons of Means[1]			
						Pollen Parents for Outcrossed Seed			
Locality		SDWT	CROSS	SDP	CROSS X SDP	A	B	C	Selfed (C)
Brownfield	F	0.7	42.3**	0.9	1.0	199a	195a	141b	125c
	df	(1,387)	(3,66)	(22,387)	(66,387)				
Allerton	F	5.4*	4.7*	0.8	0.8	264a	254a	238ab	221b
	df	(1,183)	(3,36)	(12,183)	(36,183)				

[1] Means with the same letter are not significantly different.

** $p < 0.0001$

* $p < 0.05$

Table 7. Percent increase in total pod production of outcrossed, as compared to selfed progeny of Impatiens pallida. *Data from table 6.*

Locality	*Pollen Source*		
	Between-Transects		*Within-Transect*
	A	*B*	*C*
Brownfield	59	56	13
Allerton	20	15	8

of pollen flow, the intensity of selection on characters of the pollination system can be estimated. This approach to the selective effects of plant-pollinator interactions provides the first step towards a quantitative assessment of the potential for coevolution.

Seed dispersal

Although the term "coevolution" has been used freely in the discussion of seed dispersal systems (McKey, 1975), the extent to which plants and their seed dispersers are actually coevolved is unknown. Analogous to the discussion of pollination systems, Herrera and Jordano (1981) suggest two components of plant fitness related to seed dispersal: 1) an "efficiency" or "quantity" component which is the number of seeds removed, and 2) a "quality" component which is the percent of seeds located in favorable germination sites. The evolution of mutualism specificity, and the role of coevolution between plants and dispersal agents depend in large part on the relative importance of these two components. Most research on seed dispersal systems has examined the taxonomic composition of dispersal agents for a particular plant species, and the efficiency component of dispersal success (Howe, 1977; McDiarmid et al., 1977; Howe and De Steven, 1979; Howe and Vande Kerckhove, 1979; Thompson and Willson, 1979). The quality component of dispersal success has been investigated only for ant dispersal (Culver and Beattie, 1978). The distance of seed dispersal, in relation to the spatial distribution of conspecifics, may be correlated with the quality component through germination or seedling survivorship (Janzen, 1971, 1972b; Howe and Primack, 1975; Salomonson, 1978).

Seed dispersal systems are rarely specialized, at least in comparison to what we often observe in pollination systems (Howe and Vande Kerckhove, 1979; Wheelwright and Orians, 1982). This is probably due to limited variation in the characters of plants and dispersal agents that would affect the intensity of selection on the fitness components listed here. For example, the results of field experiments conducted on two neotropical, bird-dispersed herbs indicate a tremendous advantage to dispersal into tree-fall gaps (Schemske, in press). These data suggest that any character of a seed-dispersal mutualism which increased the probability of site-specific dispersal would have a strong selective advantage. With the exception of ant-dispersal, however, seed movement beyond the canopy of a fruiting individual is probably random with respect to the spatial distribution of germination sites. Many plant species of lowland tropical forests require a treefall gap at some phase in their life history; yet few, if any, are preferentially dispersed into gaps. A long-term study of bird communities in neotropical treefalls conducted in Panama indicated no dispersal agent associated with gaps (Schemske and Brokaw, 1981).

Wheelwright and Orians (1982) suggest that "finely-tuned" mutualistic relationships are rare in seed dispersal systems, and include the following major limitations to specialization: 1) the inability of plants to provide rewards insuring site-specific dispersal, 2) small variation among dispersal agents in the quality of dispersal, 3) unpredictability of "targets" (i.e., germination sites), and 4) the advantages of utilizing a diversity of dispersal agents. To these we can add ecological constraints imposed by seed size which may limit the pool of dispersers available. The relatively broad level of habitat selection by most dispersal agents as compared to the microspatial distribution of seed germination sites certainly limits the extent to which the quality component of dispersal can be improved. It is clear that to fully understand the importance of coevolution in seed dispersal systems, future research must identify the magnitude of reciprocal selective pressures.

CONCLUSIONS

Coevolution is often assumed, but rarely demonstrated, to play an important role in plant-animal mutualisms. Mutually advantageous interactions among taxa may provide the opportunity for reciprocal selective forces, but the magnitude and direction of selection in mutualisms, and the amount of genetic variation for characters relevant to the interaction have received little attention. The examples discussed here for ant-plant, pollination, and seed dispersal systems illustrate the presence of numerous constraints on the evolution of specialized mutualisms which may also limit the potential for coevolution. In future research, ecological and genetic approaches must be combined to provide a quantitative assessment of mutualistic interactions. I propose that interactions among taxa be examined at the level of characters, with the proportionate effects on fitness determined for each set of character-state interactions. Where interactions result in joint selective effects on characters with heritable variation, coevolution is expected. As the selection intensity attributable to the interaction increases, coevolution may result in specificity among mutualists. This approach to the evolution of mutualism structure and specificity allows joint comparison of selective forces, and the responses to selection.

ACKNOWLEDGEMENTS

A.J. Beattie, W.C. Burger, L.P. Pautler and J.N. Thompson provided comments on the manuscript, and W.E. LaBerge offered numerous insights into bee biology. Robin Chase provided technical assistance. This research was supported in part by a grant from the National Science Foundation (DEB 80-22186).

LITERATURE CITED

ADDICOTT, J.F. 1981. Stability properties of 2-species models of mutualism: Simulation studies. Oecologia, 49:42-49.

BEATTIE, A.J., and D.C. CULVER. 1979. Neighborhood size in *Viola*. Evolution, 33:1225-1229.

BEATTIE, A.J. 1974. Floral evolution in *Viola*. Annals of the Missouri Botanical Garden, 61:781-793.

BECKMAN, R.L., JR., and J.M. STUCKY. 1981. Extrafloral nectaries and plant guarding in *Ipomoea pandurata* (L.) G.F.W. Mey. (Convolvulaceae). American Journal of Botany, 68:72-79.

BENTLEY, B.L. 1977a. The protective function of ants visiting the extrafloral nectaries of *Bixa orellana* L. (Bixaceae). Journal of Ecology, 65:27-38.

BENTLEY, B.L. 1977b. Extrafloral nectaries and protection by pugnacious bodyguards. Annual Review of Ecology and Systematics, 8:407-427.

BENZING, D.H. 1970. An investigation of two bromeliad myrmecophytes: *Tillandsia butzii* mez, *T. caput-medusae* E. Morren, and their ants. Bulletin of the Torrey Botanical Garden Club, 97:109-115.

BERTIN, R.I. 1982. Floral biology, hummingbird pollination

and fruit production of trumpet creeper (*Campsis radicans*, Bignoniaceae). American Journal of Botany, 69:122-134.

BIERZYCHUDEK, P. 1981. Pollinator limitation of plant reproductive effort. American Naturalist, 117:838-840.

BLOM, P.E., and W.H. CLARK. 1980. Observations of ants (Hymenoptera: Formicidae) visiting extrafloral nectaries of the barrel cactus, *Ferocactus gracilis* Gates (Cactaceae), in Baja California, Mexico. Southwestern Naturalist, 25: 181-196.

CROAT, T.B. 1978. Flora of Barro Colorado Island. Stanford, CA: Stanford University Press.

CRUDEN, R.W. 1972. Pollination biology of *Nemophila menziesii* (Hydrophyllaceae) with comments on the evolution of oligolectic bees. Evolution, 26:373-389.

CULVER, D.C., and A.J. BEATTIE. 1978. Myrmecochory in *Viola*: dynamics of seed-ant interactions in some West Virginia species. Journal of Ecology, 66:53-72.

DE NETTANCOURT, D. 1977. Incompatibility in angiosperms. New York: Springer-Verlag.

DEUTH, D. 1977. The function of extra-floral nectaries in *Aphelandra deppeana* Sch. and Cham. (Acanthaceae). Brensesia, 10:135-145.

GOH, B.S. 1979. Stability in models of mutualism. American Naturalist, 118:488-498.

GOMEZ, L. 1974. Biology of the potato-fern *Solanopteris brunei*. Brenesia, 4:37-61.

GRANT, V. 1950. The flower constancy of bees. Botanical Review, 16:379-398.

HEINRICH, B. 1975. Energetics of pollination. Annual Review of Ecology and Systematics, 6:139-170.

HEINRICH, B. 1976. The foraging specializations of individual bumblebees. Ecological Monographs, 46:105-128.

HEINRICH, B., and P.H. RAVEN. 1972. Energetics and pollination ecology. Science, 176:597-602.

HEITHAUS, E.R. 1979. Flower-feeding specialization in wild bee and wasp communities in seasonal neotropical habitats. Oecologia, 42:179-194.

HEITHAUS, E.R., D.C. CULVER, and A.J. BEATTIE. 1980. Models of some ant-plant mutualisms. American Naturalist, 116: 347-361.

HERRERA, C.M., and P. JORDANO. 1981. *Prunus mahaleb* and birds: The high-efficiency seed dispersal system of a temperate fruiting tree. Ecological Monographs, 51:203-218.

HOCKING, B. 1970. Insect associations with the swollen thorn acacias. Transactions of the Royal Entomological Society of London, 122:211-255.

HORVITZ, C.C. 1981. Analysis of how ant behaviors affect germination in a tropical myrmecochore *Calathea microcephala* (Marantaceae): Microsite selection and aril removal by neotropical ants *Odontomachus*, *Pachycondyla* and *Solenopsis* (Formicidae). Oecologia, 51:47-52.

HOWE, H.F. 1977. Bird activity and seed dispersal of a tropical wet forest tree. Ecology, 58:539-550.

HOWE, H.F., and D. DE STEVEN. 1979. Fruit production, migrant bird visitation, and seed dispersal of *Guarea glabra* in Panama. Oecologia, 39:185-196.

HOWE, H.F., and R.B. PRIMACK. 1975. Differential seed dispersal by birds of the tree *Casearia nitida* (Flacourtiaceae). Biotropica, 7:278-283.

HOWE, H.F., and G.A. VANDE KERCKHOVE. 1979. Fecundity and seed dispersal of a tropical tree. Ecology, 60:180-189.

HUXLEY, C. 1980. Symbiosis between ants and epiphytes. Biological Review, 55:321-340.

INOUYE, D.W. 1980. The effect of proboscis and corolla tube lengths on patterns and rates of flower visitation by bumblebees. Oecologia, 45:197-201.

INOUYE, D.W., and O.R. TAYLOR, JR. 1979. A temperate region plant-ant-seed predator system: Consequences of extra-floral nectar secretion by *Helianthella quinquenervis*. Ecology, 60:1-7.

JANZEN, D.H. 1966. Coevolution of mutualism between ants and acacias in Central America. Evolution, 20:249-275.

JANZEN, D.H. 1967. Interaction of the bull's horn acacia (*Acacia cornigera* L.) with an ant inhabitant (*Pseudo-myrmex ferruginea* F. Smith) in eastern Mexico. University of Kansas Science Bulletin, 47:315-458.

JANZEN, D.H. 1969. Allelopathy by myrmecophytes: The ant *Azteca* an an allelopathic agent of *Cecropica*. Ecology, 50:147-153.

JANZEN, D.H. 1971. Escape of juvenile *Dioclea megacarpa* (Leguminosae) vines from predators in a deciduous tropical forest. American Naturalist, 105:97-112.

JANZEN, D.H. 1972a. Protection of *Barteria* (Passifloraceae) by *Pachysima* ants (Pseudomyrmicinae) in a Nigerian rain forest. Ecology, 53:885-892.

JANZEN, D.H. 1972b. Escape in space by *Sterculia apetala* seeds from the bug *Dysdercus fasciatus* in a Costa Rican deciduous forest. Ecology, 53:350-361.

JANZEN, D.H. 1974. Epiphytic myrmecophytes in Sarawak: Mutualism through the feeding of plants by ants. Biotropica, 6:237-259.

JANZEN, D.H. 1980. When is it coevolution? Evolution, 34: 611-612.

JANZEN, D.H., P. DE VRIES, D.E. GLADSTONE, M.L. HIGGINS, and T.M. LEWISOHN. 1980. Self- and cross-pollination of *Encyclia cordigera* (Orchidaceae) in Santa Rosa National Park, Costa Rica. Biotropica, 12:72-74.

KAY, Q.O.N. 1976. Preferential pollination of yellow-flowered morphs of *Raphanus raphanistrum* by *Pieris* and *Eristalis* species. Nature, 261:230-232.

KEELER, K.H. 1977. The extrafloral nectaries of *Ipomoea carnea* (Convolvulaceae). American Journal of Botany, 64:112-118.

KEELER, K.H. 1981. Function of *Mentzelila nuda* (Loasaceae) postfloral nectaries in seed defense. American Journal of Botany, 68:295-299.

KERSTER, H.W., and D.A. LEVIN. 1968. Neighborhood size in *Lithospermum caroliniense*. Genetics, 60:577-587.

KLEINFELDT, S.E. 1978. Ant-gardens: The interaction of *Codonanthe crassifolia* (Gesneriaceae) and *Crematogaster longispina* (Formicidae). Ecology, 59:449-456.

KROMBEIN, K.V., P.D. HURD, JR., D.R. SMITH, and B.D. BURKS. 1979. Catalog of Hymenoptera in America north of Mexico. Washington, D.C.: Smithsonian Institution Press.

LESTON, D. 1978. A neotropical ant mosaic. Annals of the Entomological Society of America, 71: 649-653.

LEVIN, D.A. 1972. Low frequency disadvantage in the exploitation of pollinators by corolla variants in *Phlox*. American Naturalist, 106:453-460.

LEVIN, D.A. 1981. Dispersal versus gene flow in plants. Annals of the Missouri Botanical Garden, 68:233-253.

LEVIN, D.A., and H.W. KERSTER. 1974. Gene flow in seed plants. Evolutionary Biology, 7:139-220.

LEVIN, S.A., and J.D. UDOVIC. 1977. A mathematical model of coevolving populations. American Naturalist, 111:657-675.

LINSLEY, E.G. 1958. The ecology of solitary bees. Hilgardia, 27:543-599.

LINSLEY, E.G., and J.W. MACSWAIN. 1958. The significance of floral constancy among bees of the genus *Diadasia* (Hymenoptera, Anthophoridae). Evolution, 12:219-223.

LINSLEY, E.G., J.W. MACSWAIN, P.H. RAVEN, and R.W. THORP. 1973. Comparative behavior of bees and Onagraceae, V. University of California Publications in Entomology, 71:1-77.

MACIOR, L.W. 1971. Coevolution of plants and animals--systematic insights from plant-insect interactions. Taxon, 20:17-28.

MACSWAIN, J.W., P.H. RAVEN, and R.W. THORP. 1973. Comparative behavior of bees and Onagraceae, IV. University of California Publications in Entomology, 70:1-80.

MAY, R.M. 1973. Stability and complexity in model ecosystems. Princeton, New Jersey: Princeton University Press.

MAY, R.M. 1976. Models for two interacting populations. *In*: R.M. May, editor, Theoretical Ecology: Principles and applications. Philadelphia, Pennsylvania: W.B. Saunders, pages 49-70.

MAY, R.M. 1982. Mutualistic interactions among species. Nature, 296:803-804.

MCDIARMID, R.W., R.E. RICKLEFS, and M.S. FOSTER. 1977. Dispersal of *Stemmadenia donnell-smithii* (Apocynaceae) by birds. Biotropica, 9:9-25.

MCKEY, D. 1975. The ecology of coevolved seed dispersal systems. *In:* L.E. Gilbert and P. Raven, editors, Coevolution of animals and plants. Austin, Texas: University of Texas Press, pages 159-191.

MICHENER, C.D. 1954. Bees of Panama. Bulletin of the American Museum of Natural History, 100:1-175.

MICHENER, C.D. 1979. Biogeography of bees. Annals of the Missouri Botanical Garden, 66:277-347.

MOGFORD, D.J. 1974. Flower colour polymorphism in *Cirsium palustre*. 2. Pollination. Heredity, 33:257-263.

MOLDENKE, A.R. 1976. California pollination ecology and vegetation types. Phytologia, 34:305-361.

MOLDENKE, A.R. 1979a. Host-plant coevolution and the diversity of bees in relation to the flora of North America. Phytologia, 43:357-419.

MOLDENKE, A.R. 1979b. The role of host-plant selection in bee speciation processes. Phytologia, 43:433-460.

MOTTEN, A.F., D.R. CAMPBELL, D.E. ALEXANDER, and H.L. MILLER. 1981. Pollination effectiveness of specialist and generalist visitors to a North Carolina population of *Claytonia virginica*. Ecology, 62:1278-1287.

O'DOWD, D.J. 1979. Foliar nectar production and ant activity on a neotropical tree, *Ochroma pyramidale*. Oecologia, 43: 233-243.

PLEASANTS, J.M., and M. ZIMMERMAN. 1979. Patchiness in the dispersion of nectar resources: Evidence for hot and cold spots. Oecologia, 41:283-288.

PRICE, M.V., and N.M. WASER. 1979. Pollen dispersal and optimal outcrossing in *Delphinium nelsoni*. Nature, 277: 294-297.

PRIMACK, R.B., and J.A. SILANDER. 1975. Measuring the relative importance of different pollinators to plants. Nature, 255:143-144.

PYKE, G.H. 1978. Optimal foraging: Movement patterns of bumblebees between inflorescences. Theoretical Population Biology, 13:72-98.

RICKLEFS, R. 1979. Ecology. New York: Chiron Press.

RICKSON, F.R. 1980. Developmental anatomy and ultrastructure of the ant-food bodies (Beccariian Bodies) of *Macaranga triloba* and *M. hypoleua* (Euphorbiaceae). American Journal of Botany, 67:285-292.

RISCH, S., M. MCCLURE, J. VANDERMEER, and S. WALTZ. 1977. Mutualism between three species of tropical *Piper* (Piperaceae) and their ant inhabitants. American Midland Naturalist, 98:433-443.

ROBERTSON, C. 1928. Flowers and insects: List of visitors

of four hundred and fifty-three flowers. Carlinville, Ill. Lancaster, Pa.: Science Press.
ROUGHGARDEN, J. 1979. Theory of Population Genetics and Evolutionary Ecology: An Introduction. New York: Macmillan.
SALOMONSON, M.G. 1968. Adaptations for animal dispersal of one-seed juniper seeds. Oecologia, 32:333-339.
SCHAAL, B.A. 1980. Measurement of gene flow in *Lupinus texensis*. Nature, 284:450-451.
SCHEMSKE, D.W. 1977. Flowering phenology and seed set in *Claytonia virginica* (Portulacaceae). Bulletin of the Torrey Botanical Club, 104:254-263.
SCHEMSKE, D.W. 1978. Evolution of reproductive characteristics in *Impatiens* (Balsaminaceae): The significance of cleistogamy and chasmogamy. Ecology, 59:596-613.
SCHEMSKE, D.W. 1980a. The evolutionary significance of extrafloral nectar production by *Costus woodsonii* (Zingiberaceae): An experimental analysis of ant protection. Journal of Ecology, 68:959-967.
SCHEMSKE, D.W. 1980b. Evolution of floral display in the orchid *Brassavola nodosa*. Evolution, 34:489-493.
SCHEMSKE, D.W. 1980c. Floral ecology and hummingbird pollination of *Combretum farinosum* in Costa Rica. Biotropica, 12:169-181.
SCHEMSKE, D.W. 1982. Ecological correlates of a neotropical mutualism: Ant assemblages at *Costus* extrafloral nectaries. Ecology, 63:932-941.
SCHEMSKE, D.W. (In press). Breeding system and habitat effects on fitness components in three neotropical *Costus* (Zingiberaceae). Evolution.
SCHEMSKE, D.W., and N. BROKAW. 1981. Treefalls and the distribution of understory birds in a tropical forest. Ecology, 62:938-945.
SCHEMSKE, D.W., and C. FENSTER. (In press). Pollen grain interactions in a neotropical *Costus*: Effects of clump size and competitors. *In:* D.L. Mulcahy and E. Ottaviano, editors, Pollen Biology: Basic and Applied Aspects. New York: Elsevier Press.
SCHEMSKE, D.W., M.F. WILLSON, M.N. MELAMPY, L.J. MILLER, L. VERNER, K.M. SCHEMSKE, and L.B. BEST. 1978. Flowering ecology of some spring woodland herbs. Ecology, 59:351-366.
SCHMITT, J. 1980. Pollinator foraging behavior and gene dispersal in *Senecio* (Compositae). Evolution, 34:934-943.
STEPHEN, W.P., G.E. BOHART, and P.F. TORCHIO. 1969. The biology and external morphology of bees with a synopsis of the genera of Northeastern America. Agricultural Experiment Station, Corvallis, Oregon: Oregon State University.
STEPHENSON, A.G. 1979. An evolutionary examination of the floral display of *Catalpa speciosa* (Bignoniaceae).

Evolution, 33:1200-1209.
STEPHENSON, A.G. 1981. Flower and fruit abortion: Proximate causes and ultimate functions. Annual Review of Ecology and Systematics, 12:253-279.
STILES, F.G. 1981. Geographical aspects of bird-flower coevolution, with particular reference to Central America. Annals of the Missouri Botanical Garden, 68:323-351.
THOMPSON, J.N., and M.F. WILLSON. 1979. Evolution of temperate fruit/bird interactions: Phenological strategies. Evolution, 33:973-982.
THOMSON, J.D. 1978. Effects of stand composition on insect visitation in two-species mixtures of *Hieracium*. American Midland Naturalist, 100:431-440.
THOMSON, J.D., and R.C. PLOWRIGHT. 1980. Pollen carryover, nectar rewards, and pollinator behavior with special reference to *Diervilla lonicera*. Oecologia, 46:68-74.
THORP, R.W. 1969. Systematics and ecology of bees of the subgenus *Diandrena* (Hymenoptera: Andrenidae). University of California Publications in Entomology, 52:1-146.
TILMAN, D. 1978. Cherries, ants and tent caterpillars: Timing of nectar production in relation to susceptibility of caterpillars to ant predation. Ecology, 59:585-692.
VANDERMEER, J.H., and D.H. BOUCHER. 1978. Varieties of mutualistic interaction in population models. Journal of Theoretical Biology, 74:549-558.
WADDINGTON, K.D. 1980. Flight patterns of foraging bees relative to density of artificial flowers and distribution of nectar. Oecologia, 44:199-204.
WADDINGTON, K.D. 1981. Factors influencing pollen flow in bumblebee-pollinated *Delphinium virescens*. Oikos, 37: 153-159.
WASER, N.M., and M.V. PRICE. 1981. Pollinator choice and stabilizing selection for flower color in *Delphinium nelsonii*. Evolution, 35:376-390.
WHEELWRIGHT, N.T., and G.H. ORIANS. 1982. Seed dispersal by animals: Contrasts with pollen dispersal, problems of terminology, and constraints on coevolution. American Naturalist, 119:402-413.
WILLSON, M.F., and P.W. PRICE. 1977. The evolution of inflorescence size in *Asclepias* (Asclepiadaceae). Evolution, 31:495-511.
WILLSON, M.F., and D.W. SCHEMSKE. 1980. Pollinator limitation, fruit production and floral display in Pawpaw (*Asimina triloba*). Bulletin of the Torrey Botanical Club, 107:401-408.
WOLF, L.L., F.G. STILES, and F.R. HAINSWORTH. 1976. The ecological organization of a tropical highland hummingbird community. Journal of Animal Ecology, 32:349-379.
WRIGHT, S. 1946. Isolation by distance under diverse systems

of mating. Genetics, 31:39-59.
ZIMMERMAN, M. 1980. Reproduction in *Polemonium*: Competition for pollinators. Ecology, 61:497-501.
ZIMMERMAN, M. 1981. Patchiness in the dispersion of nectar resources: Probable causes. Oecologia, 49:154-157.
ZIMMERMAN, M. 1982. The effect of nectar production on neighborhood size. Oecologia, 52:104-108.

CRUSTACEAN SYMBIONTS AND THE DEFENSE OF CORALS: COEVOLUTION ON THE REEF?

Peter W. Glynn

Smithsonian Tropical Research Institute
Balboa, Republic of Panama

*A mutualism involving crustaceans and reef-building corals is examined in ecological terms and in the context of coevolution. Several xanthid crab species (*Trapezia *and* Tetralia*) and an alpheid shrimp (*Alpheus*) reside as obligate symbionts in certain corals of branching colony form. Host corals provide crustaceans with shelter from predators and with mucus as a food source; resident crustaceans protect their coral host (protected corals) against corallivores. When sea stars attempt to feed on corals sheltering such crustaceans the crustaceans drive them away by pinching, jerking, snapping and snipping off tube feet and spines, thereby preventing them from killing the host coral.*

Field experiments have shown that removing a coral's crabs and shrimps increases its likelihood of being attacked and killed by Acanthaster. *Protected corals are often abundant locally and widespread geographically, occurring in virtually all shallow, reef-coral communities in the Indo-Pacific region. The high-level aggression of the crustacean guards may be an evolutionary response directed toward host coral protection. The level of crustacean defense was greatest against* Acanthaster, *which can kill the host coral, but less intense against corallivores that do not normally cause host coral death. Resident crustacean agonism toward foreign crustaceans (experimentally introduced* Trapezia *and* Alpheus*) was less intense and often qualitatively different than against* Acanthaster, *indicating that a significant component of the crustacean's defensive behavior was directed specifically against a predatory threat. Possible responses of corals to the crustacean symbionts are the development of compact colony branching, increased production of high-quality mucus, production of "fat bodies," and a reduction in structural and* *chemical defenses.*

The pattern of defensive behavior by crustacean symbionts residing in various corals indicated that acroporid corals are weakly defended compared to pocilloporid corals. Consideration of various crustacean species that feed facultatively on coral mucus and that show some capacity for host defense are suggestive of the kinds of guard prototypes that were involved in the evolution of interdependent crustacean guard mutualisms. The evolution of host guard agonism has evidently been subject to numerous selective forces, including (a) a diverse array of corallivores, (b) predators of the crustacean guards themselves, (c) competitors for coral host resources, and (d) crustacean social interactions. A comparison of the crustacean-coral and ant-acacia (myrmecophyte) mutualisms suggests that coevolution is most apparent in the behavioral traits of the crustacean guards (frequent high-level aggression), in contrast to the ant-acacia system which shows greatest modifications in the structural attributes of host plants (swollen thorns, foliar glands, and Beltian bodies). The extensive diversification of coral families with crustacean guards during the late Tertiary does not appear to be causally related to the benefit of increased host defense.

INTRODUCTION

Coral reef communities contain numerous mutualistic species associations that are often assumed to represent coevolved relationships. The predominance of the coral reef biotope in shallow tropical seas is due in large measure to the intimate partnership between endosymbiotic algae and scleractinian corals. Corals supply essential mineral nutrients to unicellular algae (dinoflagellates) residing in their tissues, and the autotrophic algae synthesize metabolites that pass to the coral host. This mutualism has resulted in the erection of wave-resistant structures that have dominated much of the world's tropical shores over significant spans of geologic time.

Part of the fascination surrounding coral reef research stems from the variety of such intimate partnerships found in coral communities. For example, the presence of nutrient-supplying endosymbiotic algae on coral reefs has been demonstrated in numerous actiniarian and zoanthid coelenterates (Goreau et al., 1971; Muscatine and Porter, 1977), in the bivalve family Tridacnidae (Goreau et al., 1973), and in sea slugs (Trench, R.K., 1969; Trench, R.K., et al., 1969; Trench, M.E., et al., 1970). Other examples of intimate, and mutually beneficial relationships, include clownfishes that nestle unharmed among the heavily armed tentacles of giant sea anemones (Mariscal, 1966; Roughgarden, 1975; Dunn, 1981); reef fishes engaged in Batesian, Müllerian and aggressive mimicry (Losey, 1972; Russell et al., 1976); and shrimp and fish cleaners that maintain cleaning stations on coral reefs (Potts, 1973; Slobodkin and Fishelson, 1974).

In this paper I examine a relatively recently discovered mutualism involving crustaceans that live in association with, and on the products of, reef corals, and that defend their coral hosts from corallivores (coral predators). The protection these crustaceans offer is ecologically important because the corals that harbor the crustacean symbionts can be spared from predation and thus experience higher survival. (Crustacean symbionts in this paper signify crabs and shrimps involved in host coral defense.) The crustacean-coral mutualism would seem to be coevolved in the sense understood by Janzen (1980): i.e., a crustacean symbiont population induces a coral host population to produce large amounts of high-quality mucus, and the crustacean symbionts evolve a strong defensive behavior protecting the host corals from predatory attacks. Some evidence is presented here that may help to evaluate this notion. It is clear, however, that this explanation needs further testing. There are numerous aspects of the association that require further elucidation: e.g., efficacy of defense in relation to corallivore potency and abundance, and the importance of competitive and social interactions among the crustacean symbionts. Some such questions are identified and future research is suggested.

This study focuses mainly on crustacean-coral-sea star interactions in the light of results from Panama, American Samoa, and Guam. Special attention will be given to (1) the variety of species involved, (2) the efficacy of the crustacean's defensive behavior, (3) regional similarities and differences in host defense, and (4) community structure in relation to host defense. I conclude by contrasting the ant-plant and crustacean-coral mutualisms, and offer some speculative comments on the evolutionary development of crustacean coral guards.

THE INTERACTIONS

The possibility that certain crustacean symbionts associated with corals could protect their coral hosts from sea star predators was first recognized in the central and western Pacific Ocean (Pearson and Endean, 1969; Weber and Woodhead, 1970). Similar defensive behaviors were observed in the eastern Pacific (Glynn, 1976, 1980). Defense is elicited when a foraging sea star (most commonly the crown-of-thorns *Acanthaster* or the cushion sea star *Culcita*) attempts to mount the potential coral prey colony in order to extrude its stomach over the coral's living branches. Crabs and shrimps present among the branches are quickly aroused by an

---

FIGURE 1. Crustacean-coral-sea star interactions at Guam. A - The crab Trapezia cymodoce *sp. 1 (see arrow) defending its host* Pocillopora *sp. from an attacking* Acanthaster planci. *B - The shrimp* Alpheus lottini *(see arrow) defending its host* Pocillopora *sp. from an attacking* Culcita novaeguineae. *Scale bars in A and B are 2 cm long. Gun Beach, 3 April 1981, 10 meters depth.*

A

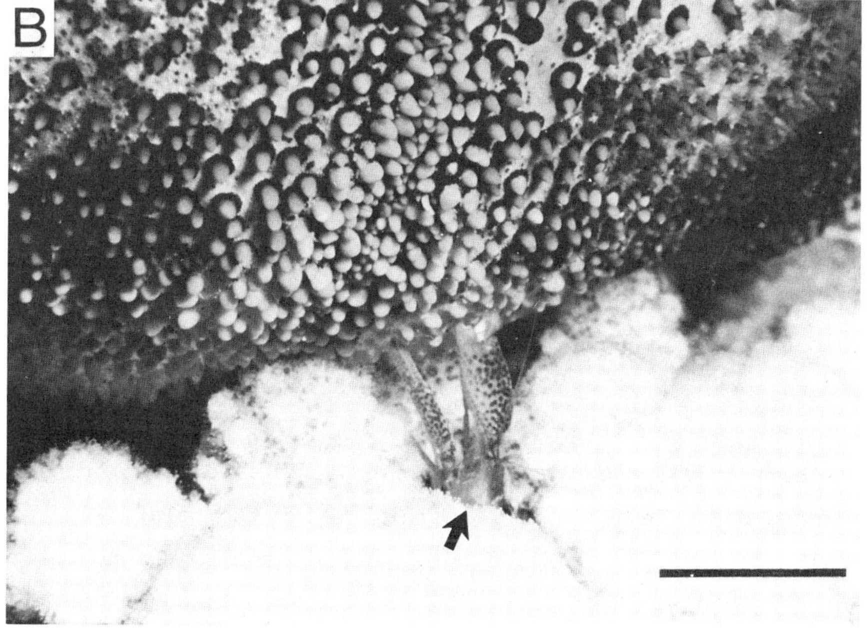
B

attacking sea star and move toward the periphery of the colony from where they harass the sea star until it moves away (crabs pinch, clip spines and tube feet, and jerk sea stars up and down; shrimps pinch and snap at sea stars, see figs. 1 and 2). Sea stars attack only the coral and thus do not directly menace the crustaceans themselves. The defending crustaceans often recognize an approaching sea star, by means of olfactory and visual cues, and respond before it even touches the coral (Glynn, 1980). Crabs and shrimps usually engage in colony defense simultaneously, and the level of response is frequently sufficient to discourage further feeding attempts.

Corals harboring xanthid crabs (several species of *Trapezia* and *Tetralia*) and alpheid shrimps (one species of *Alpheus*) have branching colony morphologies and belong to the reef-building scleractinian families Pocilloporidae and related Acroporidae (suborder Astrocoeniina). Protection of coral hosts has been observed in several species of pocilloporid (*Pocillopora* and *Stylophora*) and acroporid (*Acropora*) corals. From the initial observations on this symbiosis it became clear that the crustaceans gained much from their coral host in terms of shelter (refuge from predators) and food (coral mucus and entrapped detritus); but, since no apparent advantage was gained by the coral, the association was regarded as parasitic in nature (Knudsen, 1967). With the current understanding that predation on corals is widespread and frequent (Goreau et al., 1972; Endean, 1973; Endean and Chesher, 1973), the crustacean-coral association is presently considered to be mutually beneficial, with the relatively small but highly aggressive coral guards able to defend their hosts (Glynn, 1976; Lassig, 1977). The importance of the association is different for each partner: in a favorable habitat the coral could live without the crustaceans; however,

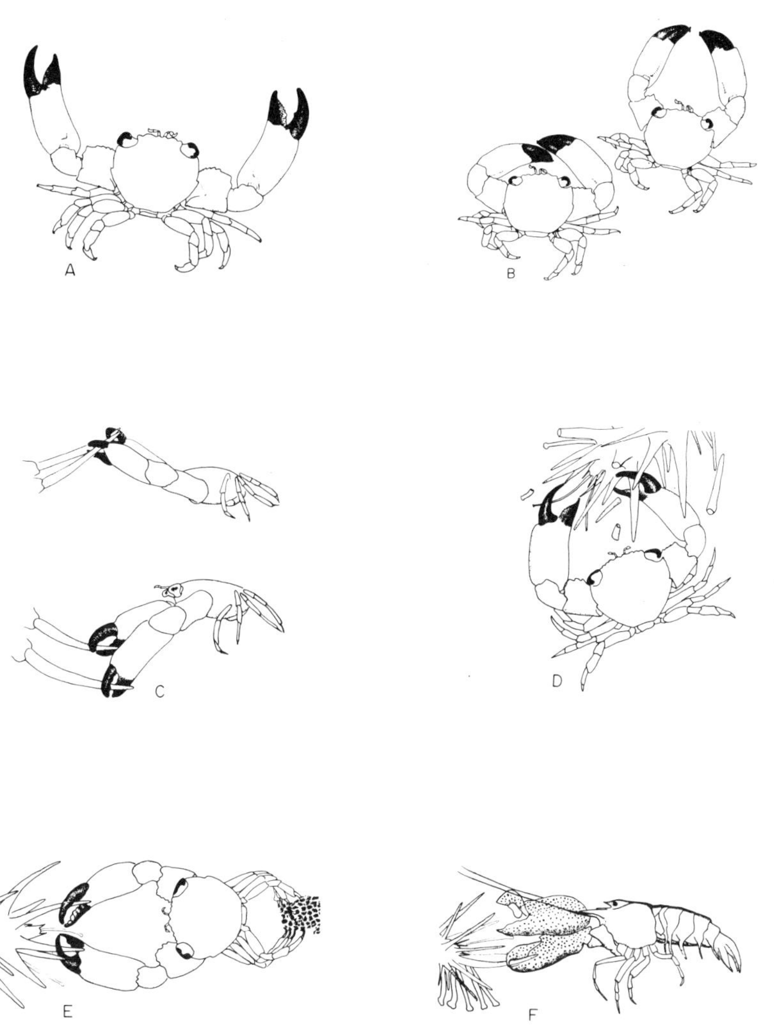

FIGURE 2. Some agonistic responses of crustacean symbionts: (A) startle display, (B) pushing, (C) up-down jerking of Acanthaster, *(D) pinching and clipping of spines and tube feet from* Acanthaster *arm tip, (E) resisting retreat of* Acanthaster, *(F) snapping in contact and pinching* Acanthaster *arm tip. A-E,* Trapezia ferruginea, *maximum carapace widths 11-12 mm, all top views except for C which is a side view. F,* Alpheus lottini, *length (from base of rostrum to posterior telson border) 20 mm, side view.*

a crustacean without its coral host would probably succumb quickly to predation or starvation.

MATERIALS AND METHODS

Field studies have been conducted on the Pacific coast of Panama (mostly from 1978-1982), on Tutuila Island, American Samoa (August 1979), and at Guam (January-June 1981) (fig. 3).

All field observations were carried out with scuba in coral habitats at depths of 1-20 m. The responses of crustacean symbionts were observed at distances of 0.5-1.5 m from the host coral, by lying or kneeling on the bottom. Individual *Acanthaster* were moved short distances (usually with a dive

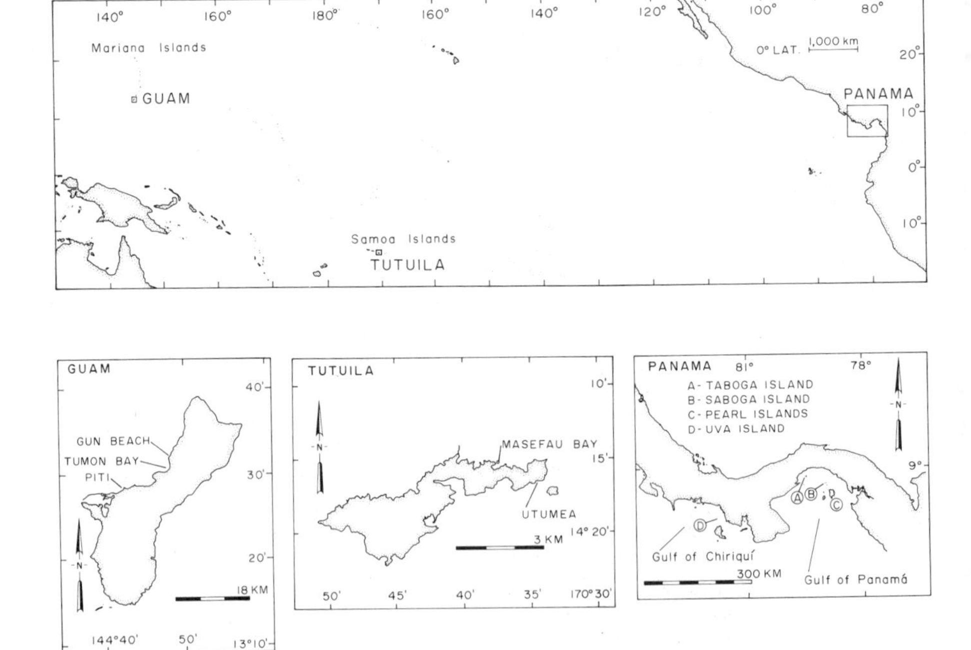

FIGURE 3. Study areas in Panama, Samoa, and Guam.

FIGURE 4. Observing and recording crustacean-sea star interactions in a simulated feeding attack (SFA). Tumon Bay, Guam, 8 June 1981, 12 meters depth.

knife) to the edges of test corals, and positioned with 2-5 arms overlapping the coral's peripheral branches. This position, a simulated feeding attack (SFA), approximated closely that assumed by naturally attacking sea stars (fig. 4). The corals tested ranged from 15 to 30 cm in diameter, and an attempt was made to select colonies of similar size. Interactions were recorded over a standard period of 3 minutes. Other test species were collected nearby and hand-placed (HP) on corals or introduced by line and rod (LAR).

The line and rod consisted of an 80-cm-long welding rod with a monofilament nylon line tied to the tip. This arrangement offered the following advantages: (1) the placement of small objects near resident crustaceans is easily controlled; (2) the retrieval of introduced objects is readily achieved without much disturbance to the coral colony; and (3) the observer is positioned at a distance from the colony. This method was used for both corallivores and crustacean symbionts.

An *Acanthaster* arm tip (2 cm long and hand-tied) was introduced (LAR) in order to equalize sizes when comparing *Acanthaster* with relatively small corallivores. The corallivores *Eucidaris* and *Trizopagurus* were hand-tied and *Jenneria* was blotted dry (upper shell surface) and cemented to the nylon line with a cyanoacrylate bonding compound. The upper carapace surfaces of crab and shrimp symbionts were also blotted dry and attached to the line by cementing. When necessary, at the end of a trial the rod tip was brushed against the legs of crustaceans to coax them away from the coral branches. Objects were always placed 5-10 cm from resident adult crabs. The crustaceans were placed in plastic bags between trials for protection against fish and octopod predators. Current direction was not considered to be important and therefore no controls were used to test its effect.

Crustacean symbionts react to the presence of corallivores both at a distance and when in contact with intruders. *Trapezia* crabs executed startle displays (fig. 2A), and lurching and shivering movements, without making physical contact. Shrimps frequently snapped at intruders from a distance. The following aggressive responses by crabs involved contact: (1) repeated pinching and clipping (spines and tube feet were the frequent targets in *Acanthaster*, fig. 2D); (2) vertical jerking movements (fig. 2C); (3) grasping spines with chelipeds and resisting retreat (fig. 2E); (4) touching and/or moving from coral onto intruding species; (5) striking; (6) pushing (fig. 2B). Responses 1-3 were vigorous (in terms of repetition and strength) high-level aggression; 4-6 were classified as low-level aggression. Shrimps also defended corals by making physical contact with intruders as follows: (1) snapping in contact with intruding species (fig. 1B); (2) pinching and clipping; (3) sometimes 1 and 2 simultaneously (fig. 2F).

All displays were recorded, but only those involving physical contact (and shrimp snapping at a distance) were scored in the analysis of corallivore-crustacean interactions. Most corallivores, even those with vision (hermit crabs and pufferfishes), did not show any apparent reaction to displays without contact. In general, each response was given a score of 1, and 1 additional tally for each second that it lasted. Each up-down jerk by *Trapezia* (fig. 2C movements equal a score of 1) and each snap by *Alpheus* was scored as 1. If *Alpheus* snapped, and pinched and clipped simultaneously, these two responses were added separately (fig. 2F response equals a score of 2). The number of responses per 3 min. refers to all interacting crustacean species, unless otherwise specified. As in quantifying the vigor of a honeybee dance (Wenner, 1971) or the defensive behavior of ants protecting plants (Bentley, 1977a), the present scoring system has its difficulties. For example, it is not known how effective a deterrent the jerking movement of crabs is, compared with the crab-holding response, or how shrimp snapping compares with each of the above.

Corals censused for crustacean symbionts were bagged *in situ*, then separated from the substrate and later transported to the laboratory. Coral size refers to maximum colony diameter unless otherwise specified. The size structure of coral populations was estimated from measurements of all colonies encountered in sample plots in habitats frequented by *Acanthaster*. Estimates of population age structures were made by relating growth rates and colony sizes. Extensive Samoa data (Mayor, 1924) were used, and some growth data from other similar areas (Tamura and Hada, 1932; Connell, 1973). Connell's (1973) analysis of growth rate versus colony size indicates that growth is essentially constant with age (after an initial rapid phase) in the species dealt with in this study. X-ray study of the main growth axes of specimens from Samoa and Guam showed uninterrupted growth (contrary to Hughes and Jackson, 1980).

The efficacy of coral host defense was tested by offering *Acanthaster*, under natural conditions, a choice between corals with and without the normal complement of crustacean symbionts. Symbionts were removed in a boat over the study site by agitating the crustaceans with a flexible wire probe. Control corals were also carried to the boat and probed with wire, but the symbionts were allowed to remain in their branches. Corals were placed on opposite sides of sea stars (touching), and prey choice was noted after 2-3 hours and again after 1 and 2 days. Preference was determined from the corals eaten. Blind (no prior knowledge of coral differences) and blind-naive (no prior knowledge of coral differences or object of experiment) testing in the laboratory in Panama gave similar results to those obtained in Guam. A blind design was not employed at Guam.

Acanthaster feeding preferences were determined primarily at Utumea (south coast) and Masefau Bay (north coast), Tutuila, American Samoa (fig. 3). These were areas where large numbers of sea stars were feeding ($\geq$ 100 individuals per hectare). Relative abundances of corals of all species were calculated from the number of links touching live and recently killed coral in chain transects 10 meters long. The relative proportions of corals eaten were calculated from the number of prey of different species killed by *Acanthaster*. Chain sample points (total = 2,866 touching live or recently killed coral at Utumea and Masefau) and number of colonies were recorded. Recently killed (bleached white) corals were assumed to have been killed by *Acanthaster*. Also, the species of corals being eaten by *Acanthaster* (n = 120 individual feeding observations) were noted.

Coral colonies selected for mucus collection were similar in size (10-15 cm, maximum diameter) and essentially free of dead surfaces (< 5% of surface area). The colonies were blotted with absorbant paper towel repeatedly for 3 min., then inverted over concave watch glasses for 15 min. The mucus-seawater secretions were collected in a Pasteur pipette and measured to the nearest 0.01 ml. Minute glass spheres were then gently applied (by blowing from a pipette) to damp skeletons from which tissues had been removed by bleaching. The spheres were removed by drying and brushing, and the coral surface area was calculated by converting weight of spheres to area.

RESULTS

Variety of Species

Although the corals and crustaceans observed in host coral defense are taxonomically restricted, several species are involved in the interaction and, with continuing study, several more may be found. Protected corals belong to the two related families Pocilloporidae and Acroporidae, and the species in these families that harbor crustacean symbionts possess a branching colony habit. Pocilloporid host corals belong to the genera *Pocillopora*, *Stylophora*, and *Seriatopora*, and acroporid host corals to *Acropora*. All crabs involved in coral defense are xanthid crabs and are members of the genera *Trapezia* and *Tetralia*. *Trapezia* crabs normally inhabit pocilloporid corals and *Tetralia* crabs inhabit acroporid corals (Patton, 1966). *Alpheus lottini* is the only known aggressive alpheid shrimp living on corals and has been found on pocilloporid corals only (Patton, 1966; and present study).

In Panama, five species of *Pocillopora* are protected by

four species of *Trapezia* and the alpheid *Alpheus lottini*. Limited observations in Samoa indicated that six pocilloporid coral taxa are defended by eight species of crab and by one species of shrimp. In Guam, nine of seventeen species of pocilloporid corals were observed to be defended by eight species of *Trapezia* and by *A. lottini*. Two species of *Pocillopora* and the pocilloporid genera *Seriatopora* (three species) and *Madracis* (two species) were not tested although they were present at Guam. *Tetralia glaberrima* defended three species of *Acropora* at Guam. It is likely that in Samoa some species of *Acropora* are also defended by *Tetralia*, but this was not explored. It is also possible that *T. glaberrima* occurs on other species of *Acropora* and that there are additional species of *Tetralia* that demonstrate some level of host coral defense.

For the eastern, central and western Pacific areas, host defense interactions were observed among thirteen coral species and eighteen crustacean species (table 1). *Acanthaster* is considered a significant threat to host corals in these areas and therefore may have been an important selective agent in the evolution of defensive behavior. Two other sea stars observed feeding on live coral in the eastern Pacific, namely *Pharia pyramidata* and *Nidorellia armata*, may cause significant coral mortality locally (Dana and Wolfson, 1970). In the central and western Pacific, the cushion sea star *Culcita novaeguineae* is also an important corallivore in certain areas (Goreau et al., 1972). Information on other corallivores is given by Robertson (1970) and Glynn (1982).

Efficacy of Crustacean Defense

Levels of Defense.

Species differences. Quantitative measures of crustacean

Table 1. Corals, crustacean symbionts, and sea star predators.

Locality	*Coral prey species*	*Crustacean symbionts*	*Sea star predators*
Panama	*Pocilloporidae*	*Xanthidae*	
	Pocillopora elegans *Dana*	Trapezia corallina *Gerstaecker*	Acanthaster planci (*Linnaeus*)
	Pocillopora damicornis (*Linnaeus*)	Trapezia formosa *Smith*	Pharia pyramidata (*Gray*)
	Pocillopora eydouxi *Milne Edwards & Haime*	Trapezia ferruginea *Latreille*	Nidorellia armata (*Gray*)
	Pocillopora capitata *Verrill*	Trapezia digitalis *Latreille*	
	Pocillopora verrucosa (*Ellis & Solander*)	*Alpheidae*	
		Alpheus lottini *Guerin*	
Samoa	*Pocilloporidae*	*Xanthidae*	
	Pocillopora elegans *Dana*	Trapezia ferruginea *Latreille*	Acanthaster planci (*Linnaeus*)
	Pocillopora damicornis (*Linnaeus*)	Trapezia digitalis *group*	Culcita novaeguineae *Müller & Troschel*
	Pocillopora eydouxi *Milne Edwards & Haime*	Trapezia aerolata *Dana*	
	Pocillopora verrucosa (*Ellis & Solander*)	Trapezia dentata *Dana*	
	Pocillopora meandrina *Dana*	Trapezia guttata *Rüppell*	
	Stylophora mordax (*Dana*)	Trapezia rufopunctata maculata *Macleay*	
		Trapezia cymodoce (*Herbst*)	
		Trapezia speciosa *Dana*	
		Alpheidae	
		Alpheus lottini *Guerin*	
Guam	*Pocilloporidae*	*Xanthidae*	
	Pocillopora elegans *Dana*	Trapezia ferruginea *Latreille*	Acanthaster planci (*Linnaeus*)
	Pocillopora damicornis (*Linnaeus*)	Trapezia digitalis *Latreille*	Culcita novaeguineae *Muller & Tröschel*
	Pocillopora eydouxi *Milne Edwards & Haime*	Trapezia rufopunctata (*Herbst*)	
	Pocillopora verrucosa (*Ellis & Solander*)	Trapezia cymodoce (*Herbst*) *sp. 1*	
	Pocillopora danae *Verrill*	Trapezia cymodoce *sp. 2*	
	Pocillopora meandrina *Dana*	Trapezia davaoensis *Ward*	
	Pocillopora setchelli *Hoffmeister*	Trapezia flavomaculata	
	Pocillopora woodjonesi *Vaughan*	Trapezia wardi *Serène*	
	Stylophora mordax (*Dana*)	Tetralia glaberrima *Herbst*	
	Acroporidae	*Alpheidae*	
	Acropora humilis (*Dana*)	Alpheus lottini *Guerin*	
	Acropora surculosa (*Dana*)		
	Acropora wardi *Verrill*		

defenses reveal that many species of pocilloporid corals receive rather similar levels of protection, regardless of either the particular suite of crustacean symbionts present or the degree of geographic separation between them. Comparisons of host coral defenses were made between coral species and between geographic areas. The latter is examined with respect to intensity of defense and its relation to predation pressure. These experiments were designed to test the hypothesis that the level of defense is causally and adaptively related to predation pressure.

The mean responses of *Trapezia* spp. (all crab species) toward *Acanthaster* in Samoa ranged from 2.3 to 4.6 per 3 min. (table 2) and was similar among the four coral species tested ($0.30 > P > 0.20$, KWT, Kruskal-Wallis test). *Alpheus* showed a significantly higher level of aggression at Samoa on *Pocillopora elegans* and on *P. damicornis* than on *P. eydouxi* or *Stylophora* ($P < 0.001$, KWT; DMCP, Dunn's multiple comparisons procedure, $\alpha = 0.20$). The low response of *Alpheus* on *Stylophora* was mainly a result of the low incidence of shrimps usually found on this coral (Patton, 1966, and present study). No *Alpheus* were found on eight colonies of *Stylophora* examined in Samoa, and 22 of 24 *Stylophora* colonies on Guam were without *A. lottini*.

The low response of *Alpheus* on *P. eydouxi* in Samoa apparently resulted mostly from the relatively large sizes of the host colonies in Samoa as compared with that in Guam. (A male-female pair of *A. lottini* is usually present in individual *Pocillopora* colonies regardless of colony size; Abele and Patton, 1976.) The median diameter of *P. eydouxi* in Samoa was nearly twice that observed on Guam (38 cm versus 20 cm, see table 8). Large colonies of *P. eydouxi* typically have an open-branching habit with most hiding places at the colony base near the attachment site. Defending *A. lottini*

Table 2. Defensive responses (number per 3 min.) of Trapezia *spp. and* Alpheus lottini *toward simulated* Acanthaster *attacks on corals.*

	Pocillopora elegans			Pocillopora damicornis			Pocillopora eydouxi			Stylophora mordax		
	$\bar{x}$	*Md*	*n*	$\bar{x}$	*Md*	*n*	$\bar{x}$	*Md*	*n*	$\bar{x}$	*Md*	*n*
Trapezia *spp.*												
Samoa	4.6	4	14	3.7	**3**	15	2.3	**2**	10	4.2	**4**	14
Guam	9.6	8	23	5.4	**3**	33	7.1	9*	27	10.2	11*	45
Panama	10.9	8	51	21.8	22.5*	16		*Uncommon*			*Absent*	
Alpheus lottini												
Samoa	8.6	7	14	13.8	6	15	0.5	**0**	10	0.2	0	14
Guam	4.6	3	23	4.1	**3**	33	6.7	6*	27	0.3	0	45
Panama	4.9	2	51	11.2	8.5*	16		*Uncommon*			*Absent*	

* *Response significantly higher than median values in boldface type among the three geographic areas (Dunn's multiple comparisons procedure, with $\alpha = 0.15$, for Kruskal-Wallis test).*

usually do not venture far beyond their shelters in contrast to defending *Trapezia* spp., which often range far onto peripheral branches. Visual inspection showed shrimps to be present on most colonies tested. A statistical comparison of shrimp abundance on all pocilloporid species is not possible because only two colonies of *P. eydouxi* were examined in Samoa; a pair of *A. lottini* was present on one colony while the second colony contained none. More extensive examination on Guam indicated that *A. lottini* occurs equally abundantly on *P. eydouxi* (mean = 0.8 shrimps per colony, n = 20 colonies sampled) and *P. damicornis* (mean = 0.9 shrimps per colony, n = 18; $P>0.05$, MWUT, Mann Whitney U test). Thus, even though shrimps are present, they may not be effective in defending all parts of large coral colonies.

Although it is impossible at present to relate the effects, for example, of a crab's jerking and a shrimp's snapping to thwarting a potential predator, it is worth noting that the frequency of the aggressive responses is similar. A statistical comparison of the median (and mean) responses, all *Pocillopora* species and localities pooled (table 2), shows the defending crustaceans to be equally active, with crabs responding about 6 times and shrimps 4.5 times per 3 min. ($P>0.05$, MWUT).

The mean responses of *Trapezia* spp. defending four pocilloporid corals at Guam ranged from 5.4 to 10.2 per 3 min. (table 2). These responses were significantly different among species ($0.01>P>0.001$, KWT); *Stylophora*, *P. elegans* and *P. eydouxi* were equally highly defended; only the latter coral was similar to the relatively low response of *Trapezia* on *P. damicornis* (DMCP, $\alpha = 0.20$). The mean responses of *A. lottini* at Guam ranged from 0.3 to 6.7 per 3 min. and also differed statistically among coral species ($P<<0.001$, KWT). The high

Table 3. Numbers of the symbiotic crab Tetralia glaberrima *found on three species of* Acropora *corals censused in Guam (Gun Beach, 5-8 meters depth, 22 and 26 March, 1981).*

Coral Species	*Number colonies sampled*	*Colony size range (LxHxW, cm^3)*	*Number of crabs per colony Median*	*Range*	*Percent colonies with adult M-F pairs*
Acropora wardi	11	360 - 5,200	2	1-2	91
Acropora humilis	10	360 - 2,600	2	2-3	100
Acropora surculosa	9	380 - 1,400	2	1-3	89

responses of *A. lottini* on *Pocillopora* spp. were significantly greater than on *Stylophora* (DMCP, $\alpha = 0.20$). This difference was due to the low abundance of the shrimp on *Stylophora*, as observed earlier in Samoa.

Pocillopora elegans and *P. damicornis* were also tested in Panama (Gulf of Chiriqui) where *Acanthaster* is present. *Trapezia* spp. showed mean responses of 10.9 and 21.8 per 3 min., and *Alpheus* 4.9 and 11.2 per 3 min. (table 2). Both *Trapezia* and *Alpheus* defended *P. damicornis* more vigorously than *P. elegans* ($P<<0.01$ in both cases, MWUT).

Only limited observations are available on the defensive behavior of *Tetralia glaberrima*, which were found occupying certain acroporid corals. Visual inspection and censusing showed crabs present in three species of *Acropora* (table 3), all colonies with closely spaced branches. Every colony sampled contained at least one adult crab, and most colonies had one mature male-female (M-F) pair (table 3). Neither *Trapezia* spp. nor *Alpheus lottini* were found on *Acropora*, and *Tetralia* was not observed on pocilloporid corals. *Tetralia* was not seen on other acroporid species with widely spaced branches when such corals were examined *in situ* on the reef

Table 4. Agonistic responses (number per 3 min.) of Tetralia glaberrima *toward* Acanthaster *introduced onto three species of* Acropora *in a simulated feeding posture (Gun Beach, Guam, 5-8 meters depth, 22 and 26 March 1981).*

Coral species	*Number colonies tested*	*Number of agonistic responses per colony:* *Mean*	*Median*	*Range*	*Percent colonies not defended*
Acropora wardi	10	2.3	0	0-10	60
Acropora humilis	10	0	0	0	100
Acropora surculosa	10	2.0	0	0-6	60

(but see Patton, 1976, and Eldredge and Kropp, in press).

Host colony defense by *Tetralia* was sporadic and weak compared with *Trapezia* (table 4). Even though all of the coral colonies tested contained crabs, 60%-100% of these were not defended. On several occasions the crabs retreated immediately from attacking *Acanthaster* to opposite sides of colonies, or approached *Acanthaster* and then quickly withdrew. When the crabs attacked *Acanthaster* they primarily pinched (repeatedly and very rapidly) the sea star's spines and tube feet. This response did not result in cutting or other apparent damage to the sea star. No aggressive acts such as jerking, resisting the sea star's retreat, or movement onto attacking sea stars were noted.

Geographic differences. If predation has been an important selective force in the evolution of coral defense, geographic variation in defense would be expected to vary in relation to corallivore abundance. One difficulty in testing this hypothesis is assessing the abundance of *Acanthaster*. Sea star abundances can fluctuate irregularly and foraging sea stars move extensively so that it is difficult to judge the long-term effect of predation in a given area. Nevertheless, available information suggests that *Acanthaster* predation decreases in intensity in the following order of localities: Panama (nonupwelling Gulf of Chiriqui), Guam, American Samoa (Tutuila) (fig. 3). A continuing study of reef areas in Panama has demonstrated relatively constant sea star population densities of 20 to 40 individuals per hectare over the past 12 years (Glynn, 1974, in press). Although Guam and Samoa may experience sporadic outbreaks involving thousands of individuals (see below), many years may elapse during which *Acanthaster* is rare and thus has little influence on coral mortality (Chesher, 1969; Weber and Woodhead, 1970; Endean and Chesher, 1973; Birkeland and Randall, 1979).

In order to identify possible geographic trends in crustacean aggression, the levels of defensive responses within particular coral host species were examined among localities. For *Trapezia* spp., most levels of defense were higher on Guam than at Samoa, and higher in Panama than on Guam or Samoa (table 2). These differences reached statistical significance for *P. damicornis* ($P \ll 0.001$, KWT), *P. eydouxi* ($P \sim 0.02$, MWUT) and *Stylophora* ($P \sim 0.005$, MWUT). *Alpheus lottini* showed higher levels of aggression on *P. damicornis* in Panama compared with Guam ($P < 0.01$, KWT) and on *P. damicornis* at Guam compared with Samoa ($P \sim 0.001$, MWUT). These results are consistent with the hypothesis that crustacean defense is positively correlated with *Acanthaster* abundance.

If high islands experience *Acanthaster* outbreaks more frequently than low islands, as proposed in Birkeland's (in press) nutrient-runoff hypothesis, then such contrasting environments would be suitable for testing coral defense in relation to predation. So far the crustacean-coral interaction has been observed only on high islands or along continental coastlines. However, some preliminary testing was done in Panama by comparing crustacean defense in the Gulf of Chiriqui (where *Acanthaster* has been present for over a decade, and possibly for several hundred years) with the Gulf of Panama, where it is absent (fig. 3). The expectation is that selection pressure for strong host defense should be relaxed in the absence of sea star corallivores and corals would thus be less effectively defended in the Gulf of Panama than in the Gulf of Chiriqui.

For *Trapezia* spp., the defensive responses toward *Acanthaster* were significantly higher at the Uva study reef, Gulf of Chiriqui (*Acanthaster* present) than at Taboga Island (Gulf of Panama); however, *Trapezia* aggression at Saboga Island (Gulf of Panama) was also significantly higher than at Taboga Island ($P<<0.001$, KWT, DMCP, $\alpha = 0.15$, table 5), but not significantly different from the Gulf of Chiriqui. The shrimp's responses to simulated *Acanthaster* attacks were significantly higher in the Gulf of Chiriqui than at either site in the Gulf of Panama ($0.01>P>0.001$, KWT).

Except for the high level of aggression shown by *Trapezia* spp. at Saboga, these results support the hypothesis that the level of host defense is positively correlated with the abundances of predators. Some potentially complicating factors, however, must be discussed. First, the comparisons were made in two very different environments, one nonupwelling (Chiriqui) and the other upwelling (Panama), and therefore the effects of the sea star can not be easily disassociated from other influences. An example of how lowered seawater temperature affects the frequency of crustacean aggression was observed during an upwelling episode at Taboga Island (15 February 1980). At that time the sea surface temperature was 20°C (6° to 8°C below the usual value), and mean defensive responses of crustaceans on *P. damicornis* were only 1.5 per 3 min. (median = 1) for *Trapezia* spp. and 1.0 per 3 min (median

Table 5. Comparisons of crustacean defensive responses (number per 3 min.) in Panama in areas with or without Acanthaster. *All test corals were* Pocillopora damicornis.

	Acanthaster *absent*		Acanthaster *present*
	Taboga Island (Aug. 1980)	*Saboga Island (Nov. 1980)*	*Uva study reef (Mar. 1982)*
Trapezia *spp.*			
Median response	8	*18*	*22.5**
0.95 conf. lim. of median	*3-14*	*0-43*	*15-31*
Number of colonies tested	*27*	*14*	*16*
Alpheus lottini			
Median response	3	2	*8.5**
0.95 conf. lim. of median	*0-6*	*0-8*	*5-19*
Number of colonies tested	*27*	*14*	*16*

* *Response significantly higher than median values in boldface type (Dunn's multiple comparisons procedure, with α = 0.15, for Kruskal-Wallis test).*

= 0) for *Alpheus*. These levels of defense were significantly lower than those observed at Taboga (table 5) during the non-upwelling season (*Trapezia*, $P<0.0001$, MWUT; *Alpheus*, $P<0.0005$, MWUT).

Another factor that might affect crustacean aggression is the difference in biotic composition of *Pocillopora* colonies in the two gulfs. Colonies of equal size contain a higher diversity of decapod crustacean species in the Gulf of Panama than in the Gulf of Chiriqui (Abele, 1976). Coral colonies in the two gulfs contain equal numbers of individuals, but the abundances of juvenile and subadult *Trapezia ferruginea* is greater in Chiriqui than in Panama. Thus, competition between obligate mucus feeders could also be greater in Chiriqui and lead to higher levels of defense.

The extent of genetic interchange between organisms in the two gulfs was presumed to be very low, but this may not be the case. It is possible that the gyre with counterclockwise circulation centered in the Panama Bight (Forsbergh, 1969) regularly conveys plankton-borne larvae between the gulfs of Chiriqui and Panama. Also, Saboga Island is located in the open Gulf of Panama and may thus be subject to more water exchange with Chiriqui than is Taboga.

The tendency for crustacean defense to be higher on Guam than at Samoa (table 2) deserves further comparisons to the abundances of *Acanthaster* in these two areas. In recent years (since the 1960s), Guam has experienced two *Acanthaster* outbreaks (1968-1970 and 1979-1981; Chesher, 1969; Birkeland and Randall, 1979) and Samoa (Tutuila) one outbreak (1977-1979; Birkeland and Randall, 1979). Finally, one would expect Panamanian coral defense, in areas where *Acanthaster* is presumably always present, to be even higher than on high west and central Pacific islands subject to irregular *Acanthaster* predation. This was observed only for *Pocillopora damicornis* (table 2).

Mortality Rates of Protected and Unprotected Corals. Not all colonies of protected coral species are always defended. The proportion of pocilloporid colonies with crustacean symbionts (excluding *Stylophora* where the shrimp is usually absent) not defended over 3-minute observation periods was usually 20% or more in 18 cases (table 6). This occasionally can be sufficient time for *Acanthaster* to mount and extrude its stomach over much of a coral colony. Once *Acanthaster* begins to feed on a coral, the crustaceans' attack rate may decline and at least part of the colony is killed. The

Table 6. Proportion of corals not defended by crustacean symbionts over 3-minute time periods.

	Pocillopora elegans		Pocillopora damicornis		Pocillopora eydouxi		Stylophora mordax	
	%	*n*	*%*	*n*	*%*	*n*	*%*	*n*
Trapezia *spp.*								
Samoa	*21.4*	*14*	*20.0*	*15*	*20.0*	*10*	*14.3*	*14*
Guam	*8.7*	*23*	*30.3*	*33*	*22.2*	*27*	*15.6*	*45*
Panama	*5.9*	*51*	*0*	*16*	*Uncommon*		*Absent*	
Alpheus lottini								
Samoa	*35.7*	*14*	*26.7*	*15*	*80.0*	*10*	*78.6*	*14*
Guam	*43.5*	*23*	*27.3*	*33*	*29.6*	*27*	*93.3*	*45*
Panama	*33.3*	*51*	*6.2*	*16*	*Uncommon*		*Absent*	

majority of the pocilloporid prey consumed by *Acanthaster* in Panama represents broken branches with few and often small crustaceans or none (Glynn, 1976). As in plants protected by ants (Bentley, 1977b), potential predators are most effectively averted if attacked vigorously while exploring or just beginning to feed on a prey.

To test the competency of crustacean defense under natural field conditions, coral prey with and without crustaceans were offered to *Acanthaster* at Guam. These observations were made on sea stars that were part of an actively foraging swarm in excess of 100 individuals. For both coral species tested, colonies stripped of their crustacean symbionts were eaten significantly more frequently than control (unstripped) colonies (table 7). Thus, the probability of available unprotected corals being eaten in an area subject to intense predation was about 0.70 to 0.80 over a 2-3-day period. These results in Guam are similar to laboratory preference tests reported from Panama. *Acanthaster*, when offered *P. damicornis* with and without crustacean symbionts, fed on 85% of the unprotected colonies (Glynn, 1976). Additional

Table 7. Acanthaster *prey choice of pocilloporid corals with normal complement of crustacean symbionts and corals stripped of crustacean symbionts (Gun Beach, Guam, 8-12 meters depth, 28-30 April 1981).*

Coral species	*Time from start of observations*	*Number stripped corals eaten*	*Percent stripped corals eaten*	*Number control corals eaten*	*Number failures to feed on corals offered*	x^2 *statist*
Pocillopora eydouxi						
(19 pairs)	*4-6 hours*	*13*	*81.2*	*3*	*22*	*6.25***
	1 day	*17*	*70.8*	*7*	*14*	*4.16***
	2 days	*18*	*72.0*	*7*	*13*	*5.66***
Stylophora mordax						
(10 pairs day 1;	*1 day*	*9*	*81.8*	*2*	*9*	*4.58***
20 pairs day 2)	*2 days*	*17*	*77.3*	*5*	*18*	*10.41***

* *Yates' correction applied*
** *P<0.05*
*** *P<0.02*

examples of the alteration of mortality rates among different coral species combinations, in the presence or absence of crustacean symbionts, are presented by Glynn (1976). Janzen (1966a) found that in 10 months unprotected acacias suffered 55%-58% mortality as compared with 28% mortality for protected plants. Although the comparison of the mortality rates in marine and terrestrial systems is not possible, it is apparent that corals and plants harboring aggressive arthropods enjoy a significant selective advantage over their congeners without this form of protection.

Effect on Coral Community Structure

Species Diversity. Preferential feeding by *Acanthaster* on relatively uncommon coral species in Panama tends to lower local species diversity or richness (Glynn, 1974, 1976; Glynn et al., in press). Whole *Pocillopora* spp. colonies harboring crustacean symbionts are eaten less frequently by *Acanthaster* than several nonpocilloporid species without protective symbionts (in contrast to Porter, 1972, 1974). Pocilloporid corals are strong competitors for a variety of reasons (e.g., high growth rates, overtopping growth habit, frequent asexual propagation, development of sweeper tentacles -- see Porter, 1974; Glynn, 1976; and Wellington, 1980) and benefit even further in coral communities where *Acanthaster* is an active forager. Under such positive effects of competition and predation, pocilloporid corals frequently predominate in Panama in all reef zones.

In contrast to the above, more recent observations in Samoa and Guam show that in reef coral communities there moderate grazing by *Acanthaster* often has a diversifying effect (Glynn and Colgan, unpublished observations). The influence that coral guards can have on community diversity depends in large part on the local relative abundances of preferred and nonpreferred corals. For example, if unguarded palatable corals are abundant, and unpalatable and guarded corals rare (less abundant), then predation will tend to increase diversity. On the other hand, if guarded corals are abundant and unguarded, palatable corals rare, then predation will tend to lower diversity.

Age Structure. Some preliminary observations suggest that the sizes, and thus age structures, of many corals are affected by the intensity of *Acanthaster* predation (Glynn et al., ms.). It was suggested above that coral communities in Guam have been subject to higher predation rates than in American Samoa (Tutuila). Since 1967, Guam has experienced two *Acanthaster* outbreaks (Chesher, 1969; Glynn and Colgan,

Table 8. Colony size differences of three living pocilloporid coral populations sampled at various sites in Guam (February-June 1981) and Samoa (August 1979). Reef zones sampled were 5-20 meters in depth, in areas frequented by Acanthaster.

	Guam			Samoa		
	Median diameter (cm)*	*0.95 conf. lim. of median*	*Number of colonies*	*Median diameter* (cm)*	*0.95 conf. lim. of median*	*Number of colonies*
Pocillopora elegans	22	20-24	133	28	25-29	110
Pocillopora eydouxi	20	18-23	42	38	32-45	41
Stylophora mordax	16	15-19	344	28	24-34	42

* *Maximum colony diameter*

unpub. obs.) and Tutuila only one (Weber and Woodhead, 1970; Vine, 1970; Birkeland and Randall, 1979). Assuming more frequent episodes of intense predation on corals at Guam than at Samoa, one may expect that coral populations would be smaller and more youthful at Guam.

The three pocilloporid species sampled in Samoa had significantly larger median colony sizes than in Guam ($P \ll 0.001$, MWUT, table 8). I suspect that these differences are due mainly to a lower predator-induced mortality in Samoa than at Guam, caused by the numbers of feeding sea stars and rates of local coral prey depletion. Some reef areas that experienced heavy predation by sea stars contained chiefly large *Pocillopora* spp. colonies (e.g., Randall, 1973), young encrusting corals of other species and numerous large colonies of species (*Porites, Synaraea, Galaxea, Millepora*) that are not commonly eaten (due perhaps to potent nematocysts or noxious secondary metabolites).

The proportion of coral species eaten by *Acanthaster* in American Samoa were tentatively compared with estimates of the

Table 9. Electivity indices and respective significance levels for proportions of corals eaten and not eaten by Acanthaster *(Samoa, August 1979).*

Coral species	*Number of colonies*	*Electivity index**	*Chi-square test*** x^2	P
Stylophora mordax	32	**0.26*****	4.13	<0.05
Acropora corymbosa *(Lamarck)*	200	**0.21**	18.91	<0.001
Acropora hyacinthus *(Dana)*	171	0.20	1.50	>0.20
Acropora leptocyathus	35	-0.03	1.03	>0.30
Pocillopora *spp.*	151	-0.17	2.15	>0.10
Porites andrewsi *Vaughan*	32	**-0.62**	4.13	<0.05
Galaxea fascicularis *(Linnaeus)*	40	**-0.75**	6.81	<0.01
Porites (Synaraea) iwayamaensis *Eguchi*	37	**-0.88**	5.30	<0.05
Porites lobata *Dana*	8	-1.00	1.53	>0.20

* *Ivlev's electivity index,* $E = (r_i - p_i)/(r_i + p_i)$, *where* r_i *is the proportion of coral eaten and* p_i *is the proportion of coral available in the environment.*

** *Yate's correction for continuity applied.*

*** *Boldface type indicates* $P \leq 0.05$ *in* x^2 *testing.*

coral's ages and relative palatabilities. Coral species preferred by *Acanthaster* (table 9), based on the proportions eaten in relation to their relative abundances, tended to have youthful age structures (2-10 years; fig. 5). Such corals included *Acropora*, only weakly defended by *Tetralia*, and *Stylophora*, usually defended by *Trapezia* only. On the other hand, nonpreferred corals suffered lower mortality and their populations contained older colonies (6-100 years). This trend, with $\tau = -0.48$, is significant ($P \sim 0.003$, Kendall rank correlation). (It is also significant when the outlier

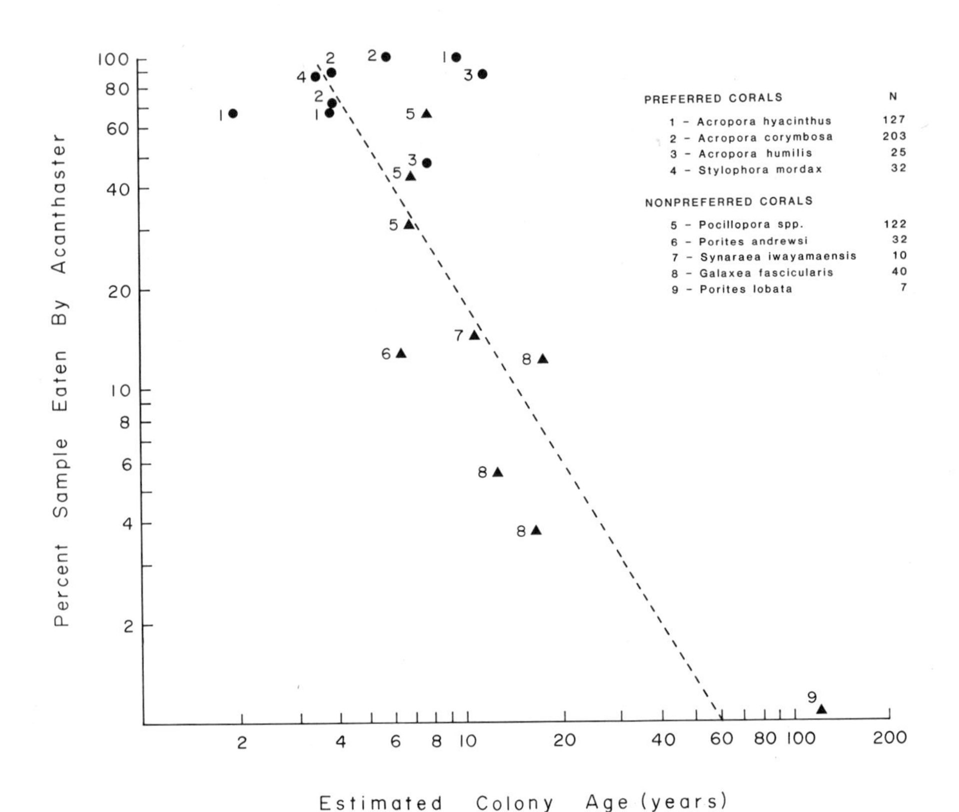

FIGURE 5. Relationship between coral species eaten by Acanthaster *and estimates of their respective ages. Preferred (circles) and nonpreferred (triangles) species are indicated (see table 9). Total number of colonies sampled is noted under* N *for each species. Regression line fitted after method of Brown and Mood (from Daniel, 1978).*

Porites lobata is omitted from the analysis; $\tau = -0.42$ $P \sim 0.009$.) I do not believe that this relationship is seriously affected by possible species differences in determinate growth. A few *Acropora* and *Pocillopora* colonies in Samoa had attained large sizes with estimated ages of 15 to 20 years. Moreover, Connell (1973) noted a median age of 50 years for an *Acropora* species in Australia.

Although observations in four of the nine Samoan data sets used to calculate the electivity indices (E) were not significant, more extensive sampling on Guam tended to confirm these results. The only exception was the coral *Stylophora*, which was preferred at Samoa (table 9) but avoided in Guam ($E = -0.52$, $P \sim 0.006$, $n = 39$; Glynn and Colgan, unpub. data). This difference shows a logical correspondence with the levels of defensive behavior of *Trapezia*, which in Guam was over two times that observed in Samoa (table 2). However, *Pocillopora eydouxi* was also weakly defended in Samoa, compared with other Samoan *Pocillopora* (table 2), but showed no indication of an expected higher mortality ($P > 0.05$, $X^2 = 3.19$, $n = 145$ colonies). These patterns of predation vis à vis symbiont defense are certainly influenced by other factors and are probably regulated by variations in the availability of preferred and nonpreferred prey as local food sources are depleted (Ormond et al., 1976).

Intensity of Aggression Toward Potential Predators

If the defensive behavior of crustacean symbionts is an adaptive response effecting increased protection of an essential resource, then it would seem reasonable to expect the symbionts to react most strongly toward the potentially most significant predators. High levels of aggression, and thus host protection, should increase the fitness of both coral hosts and their crustacean symbionts. A variety of known corallivores of varying potency were tested by bringing them into contact with host corals in simulated feeding attacks, in order to assess the levels of aggression of the resident crustacean symbionts. There are at least three difficulties with this kind of test: (1) it assumes that the effects (selective regimes) of corallivores today and in the past are

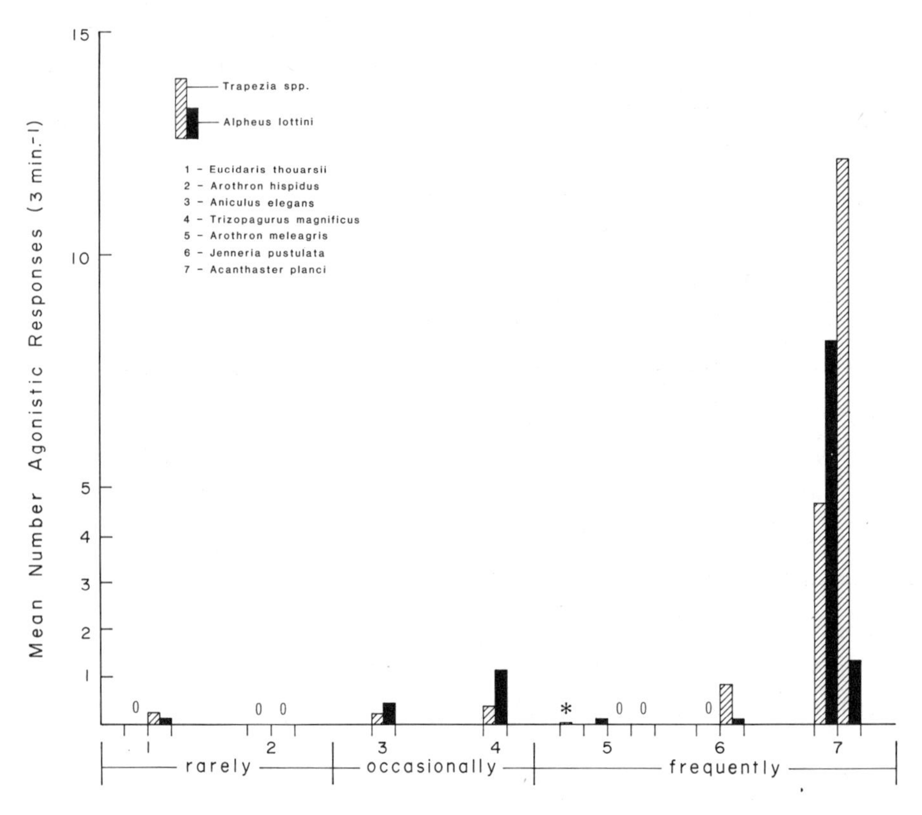

FIGURE 6. Mean aggressive responses of crustacean symbionts toward a variety of species in relation to their predatory threat to coral host. Zeros signify trials with no responses (see table 10); asterisk denotes response of less than 0.1.

approximately equal; (2) spatially and temporally, the variance of the responses can be high; (3) host defense may not be graduated in intensity.

Four corallivores that feed rarely or in small amounts on pocilloporid corals elicited relatively low levels of agonism (fig. 6, nos. 1-4; table 10). In all but one instance (*Trizopagurus*, no. 4), the mean combined number of aggressive responses was less than 1 per 3 minutes. *Eucidaris thouarsii*,

Table 10. Agonistic responses of crustacean symbionts on Pocillopora *toward a variety of animal species (and a coral fragment). Comments on the potential harm of the test subjects are given. All observations were made on* Pocillopora elegans *at the Uva Island, Panama, study reef unless noted otherwise.*

Introduced species	Protocol*	N*	Mean number (and range) of agonistic responses (per 3 min.) Trapezia**	Alpheus**	Potential harm
SEA STARS					
Acanthaster planci	SFA, Jan 1980	23	4.7 (1-14)	8.2 (0-22)	Death of entire host colony.
	SFA, Dec 1980	18	17.6 (0-54)	2.7 (0-17)	
	SFA, Dec 1981	10	12.2 (7-23)	1.3 (0-9)	
	SFA,*** Mar 1982	16	21.8 (8-39)	11.2 (0-33)	
	SFA,*** Nov 1980 Pearl Islands	14	21.7 (0-47)	4.3 (0-10)	
	SFA,*** Aug 1980 Taboga Island	27	9.2 (0-32)	3.6 (0-9)	
	LAR, arm tip, Dec 1981	36	4.3 (0-23)	1.3 (0-9)	
Oreaster occidentalis	SFA, Jan 1980	8	1.0 (0-4)	0.6 (0-4)	None observed in field; feeds on sponges; feeds on Pocillopora spp. in laboratory when starved.
	SFA, Dec 1981	10	2.8 (0-8)	0.4 (0-4)	
SEA URCHIN					
Eucidaris thouarsii	HP, Jan 1980	14	0	0	Rarely grazes on coral branches in captivity (Glynn et al., 1979; Glynn and Wellington, in press).
	LAR, Dec 1981	10	0.2 (0-2)	0.1 (0-1)	
GASTROPOD					
Jenneria pustulata	HP, Jan 1980	7	0	0	Usually partial death of host colony; occasionally death of entire colony (Glynn et al., 1972).
	LAR, Dec 1981	20	0.8 (0-4)	0.1 (0-2)	
XANTHID CRAB					
Trapezia corallina	HP, Jan 1980	10	0.4 (0-4)	0.5 (0-5)	Competitor for food, shelter, and mate (Castro, 1976).
	LAR, Dec 1981	36	1.3 (0-5)	0.2 (0-2)	
HERMIT CRABS					
Trizopagurus magnificus	LAR, Dec 1981	13	0.4 (0-2)	1.2 (0-12)	Occasionally grazes on coral and shelters among coral branches (Glynn et al., 1972).
Aniculus elegans	HP, Jan 1980	8	0.2 (0-2)	0.4 (0-3)	Occasionally grazes on coral and shelters among coral branches (Glynn et al., 1972).
PUFFERFISHES					
Arothron meleagris:					
Yellow phase	SFA, Jan 1980	7	0	0.1 (0-1)	Crops branch tips of coral host (Glynn et al., 1972).
	SFA, Dec 1981	10	0	0	
Blue-spotted phase	SFA, Jan 1980	8	0	0	
	SFA, Dec 1981	10	<0.1 (0-1)	0	
Arothron hispidus	SFA, Jan 1980	8	0	0	Rarely crops coral branch tips (personal observations).
	SFA, Dec 1981	10	0	0	
Diodon holacanthus	SFA, Dec 1981	10	0	0	Predator of Jenneria and other molluscs in coral host.
CORAL FRAGMENT					
Bleached Pocillopora branch (2 cm)	LAR, Dec 1981	10	<0.1 (0-1)	0	Rarely temporary obstruction of polyps.

* *N = number of observations; SFA = simulated feeding attack; LAR = line and rod; HP = hand-placed on coral.*

** *Of the four symbiotic crab species present,* Trapezia corallina *engaged most frequently in coral host defense;* Alpheus lottini *was the only shrimp showing an agonistic response.*

*** *Coral host was* Pocillopora damicornis.

the club-spined sea urchin, elicited a response only when introduced by line and rod (LAR). *Arothron hispidus*, a pufferfish, failed to arouse the crustacean symbionts. The two hermit crabs, *Aniculus elegans* and *Trizopagurus magnificus*, elicited responses with *Alpheus* showing about twice the activity of *Trapezia*. In Panama, *Eucidaris* feeds on live coral (in captivity) only after having been deprived of other prey (Glynn et al., 1979). The pufferfish occasionally grazes on the coral *Psammocora stellata* (without crustacean symbionts), and rarely on *Pocillopora*. The hermit crabs feed on *Pocillopora* in moderate amounts in captivity, but their impact on pocilloporid corals in the field is still largely unknown.

Of the three corallivores that feed almost exclusively on pocilloporid corals (fig. 6, nos. 5-7; table 10), the pufferfish *Arothron meleagris* and the gastropod *Jenneria pustulata* provoked only weak defensive responses. In all trials the mean responses of crabs and shrimp to the fish and gastropod corallivores was less than 1 per 3 min. *Acanthaster planci*, the crown-of-thorns sea star, elicited high levels of defense. The mean numbers of responses of *Trapezia* to *Acanthaster* were 4.7 and 12.2 per 3 min.; and of *Alpheus* to *Acanthaster*, 1.3 and 8.2 per 3 min. (Other data sets involving different times, host coral species, protocols, and localities are presented in table 10). Even the 2-cm arm tip of *Acanthaster* elicited 4.3 and 1.3 responses per 3 min. from *Trapezia* and *Alpheus* respectively.

Arothron meleagris feeds on *Pocillopora* by snipping off the 1-2 cm branch tips. These pufferfish are capable of cropping large amounts of coral, but seldom, if ever, kill entire colonies. While feeding, their time in contact with the coral colony is brief -- a branch is snipped from the colony in less than 1 sec. -- thus offering a distant and fleeting target to crustacean symbionts. *Jenneria* feeds on the tissues of

Pocillopora and typically grazes around the base of the colony; it usually grazes among the peripheral branches only after the basal branches have been stripped. Whole *Pocillopora* colonies are occasionally killed when attacked by large feeding aggregations of this snail (≥ 50 individuals). *Acanthaster* is by far the potentially most important corallivore. The sea star can kill an entire *Pocillopora* colony (up to 15 cm in diameter) in a single bout (2-3 hours). Although *Acanthaster*, along the deep reef edge at the Uva Island reef, feeds predominantly on small broken branches of pocilloporid corals and other non-pocilloporid species (Glynn, 1974, 1976), whole *Pocillopora* colonies are frequently approached, and sometimes eaten.

Summarizing the crustacean's responses to corallivores, both *Trapezia* and *Alpheus* reacted more frequently to *Acanthaster* than to any of the other corallivores tested ($P \ll 0.001$, KWT, $n = 137$, DMCP, $\alpha = 0.25$). Even the defensive responses elicited by the arm tip of *Acanthaster* tended to be high. The responses of *Trapezia* spp. to the arm tip was greater than their responses to any of the other corallivores ($P \ll 0.001$, KWT, $n = 140$, DMCP, $\alpha = 0.25$). The mean response of *Alpheus* was significantly greater to the arm tip of *Acanthaster* when contrasted with the pufferfish *A. meleagris* ($0.01 > P > 0.001$, KWT, $n = 140$, DMCP, $\alpha = 0.15$).

The aggressive activities of *Trapezia* and *Alpheus* are sometimes markedly different under apparently similar conditions. For example, the mean number of responses by *Trapezia* toward *Acanthaster* in January 1980 (4.7 per 3 min.) was significantly lower than in December 1981 (12.2 per 3 min.; $P \ll 0.0001$, MWUT); however, *Alpheus* was significantly more active in January 1980 (8.2 responses per 3 min.) than in December 1981 (1.3 responses per 3 min.; $P < 0.0005$, MWUT). This difference is inexplicable in terms of such obvious factors as season, time of day, tidal state, and water clarity.

If the combined responses of crabs and shrimps are compared between years, i.e., the combined individual values making up the means of 12.9 (4.7 + 8.2, 1980) and 13.5 (12.2 + 1.3, 1981) responses per 3 min., it is seen that the overall levels of aggression were not different ($P \sim 0.32$, MWUT). This result raises the question: Do crabs and shrimps defend their host corals in a compensatory manner?

Five data sets involving crab and shrimp aggression toward *Acanthaster* were examined for possible joint (compensatory or combined) defensive responses. That is, if shrimp defense is low, will crab defense be high (compensatory) or are shrimp and crab defenses reinforcing (combined). In only one case was a significant correlation found between the frequency of defensive responses in crabs and shrimps, and this was positive (Uva study reef, January 1980; Kendall's $\tau = 0.48$, $P \sim 0.0007$), suggesting a joint reinforcing response. However, in four instances, no significant correlations were evident (Uva study reef: December 1980, $\tau = 0.12$, $P \sim 0.24$; December 1981, $\tau = -0.04$, $P \sim 0.50$; March 1982, $\tau = 0.08$, $P \sim 0.33$; Saboga reef: November 1980, $\tau = 0.25$, $P \sim 0.13$). Thus, it is not obvious from these results that any sort of compensatory or reinforcing agonism is involved in host coral defense. Definitive testing, however, will require detailed information on the relative abundances and relations between crab and shrimp guard species for each host colony.

Intensity of Aggression Toward Potential Competitors

To test levels of aggression toward potential competitors, foreign crustacean symbionts were introduced individually (LAR) into pocilloporid corals already possessing a natural complement of symbionts. In Guam, only crustaceans (crabs and shrimps) were introduced into corals but, in Panama, crustaceans (crabs) and corallivores (*Acanthaster* arm tip and *Jenneria*) were introduced alternately into corals. The reason for the procedure in Panama was to test crustacean defense against competitors and predators simultaneously in order to minimize the variability between colonies.

Resident crustaceans exhibited relatively low levels of aggression toward three species of foreign crustacean symbionts at Guam. The mean responses of *Trapezia davaoensis* toward foreign *Trapezia* spp. and *A. lottini* ranged from 0.4 to 1.7 per 3 min. (table 11). The defensive response of *T. davaoensis* toward *Acanthaster* (in the same population of *P. damicornis*, but in different colonies and at different times) was 4 to 14 times the response toward introduced crustaceans (table 2). The mean responses of resident *Trapezia* toward introduced *T. davaoensis* (0.4 per 3 min.) and *T. cymodoce* 1 (1.3 per 3 min.) were significantly lower than toward *Acanthaster* (5.4 per 3 min., table 2) ($P<0.05$, MWUT), but marginally insignificant for *Alpheus* (1.7 per 3 min.) compared with *Acanthaster* ($P = 0.057$, MWUT). The mean levels of aggression of *Alpheus* toward foreign *Trapezia* spp. and *A. lottini* ranged from 0.3 to 1.5 responses per 3 min. (table 11). The levels of aggression of the defending shrimps were 3 to 14 times greater toward *Acanthaster* compared with the introduced crustaceans (4.1 per 3 min., table 2). *Alpheus lottini* aggression toward potential crustacean competitors and the corallivore *Acanthaster* was significantly greater toward *Acanthaster* in all three comparisons ($P<0.05$, MWUT).

The apparent lower rates of intraspecific versus interspecific aggression among crustacean symbionts, i.e., 0.4 responses per 3 min. for *T. davaoensis* / *T. davaoensis*, and 0.3 responses per 3 min. for *A. lottini* / *A. lottini* (table 11), were not significantly different for the indicated

Table 11. Agonistic responses (number per 3 min.) of resident species Trapezia davaoensis *and* Alpheus lottini *toward three introduced crustacean symbiont species. All host corals tested were* Pocillopora damicornis *(Piti, Guam, 1 meter depth, 7 June 1981).*

Introduced symbionts:	Trapezia davaoensis		Trapezia cymodoce 1		Alpheus lottini	
Resident species:	T. davaoensis	A. lottini	T. davaoensis	A. lottini	T. davaoensis	A. lottini
Mean response	0.4	1.2	1.3	1.5	1.7	0.3
Median response	0	1	0	0	2	0
Range	0-2	0-4	0-5	0-7	0-6	0-3
Number of colonies tested	10	10	10	10	9	9

Table 12. Agonistic responses (number per 3 min.) of resident Trapezia *spp. and* Alpheus lottini *toward foreign* T. ferruginea *and arm tip of* Acanthaster *introduced alternately. All host corals tested were* Pocillopora elegans *(Uva Island study reef, Panama, December 1981).*

Introduced species:	Trapezia ferruginea		Acanthaster *arm tip*	
Resident species:	Trapezia *spp.*	A. lottini	Trapezia *spp.*	A. lottini
Mean response ($\pm s_{\bar{x}}$)	2.2 ± 0.5	0.2 ± 0.1	7.8 ± 2.6	1.8 ± 0.6
Median response	2 (0-3)*	0	6 (3-8)	0
Range	0-10	0-2	0-46	0-17
Number of colonies tested	40	40	40	40

* *0.95 confidence limits of median.*

comparisons (P>0.05 in both tests, KWT). On the other hand, Preston (1973) and Abele (in press) reported more frequent intraspecific than interspecific aggression in *Trapezia* and in *A. lottini*.

Trapezia and arm tips of *Acanthaster*, introduced alternately (first one species, then the other) into corals in Panama, elicited low and high levels of aggression respectively, similar to the responses observed in Guam (table 12). For example, the mean response of resident *Trapezia* spp. toward *Acanthaster* was over three times the response of resident *Trapezia* spp. to foreign *Trapezia ferruginea* (7.8 versus 2.2 responses per 3 min., P<<0.001, WSRT, Wilcoxon matched pairs signed-ranks test). Likewise, the mean response of resident *Alpheus* toward *Acanthaster* (1.8 per 3 min.) was nine times the shrimp's response toward a foreign crab (0.2 per 3 min.;

Table 13. Agonistic responses (number per 3 min.) of resident Trapezia *spp. and* Alpheus lottini *toward foreign* T. corallina *and* Jenneria pustulata *introduced alternately. All host corals tested were* Pocillopora elegans *(Uva Island study reef, Panama, December 1981).*

Introduced species:	Trapezia corallina		Jenneria pustulata	
Resident species:	Trapezia *spp.*	A. lottini	Trapezia *spp.*	A. lottini
Mean response ($\pm s_{\bar{x}}$)	*2.2 ± 0.6*	*0.3 ± 0.2*	*1.6 ± 0.6*	*0.1 ± 0.1*
Median response	*1.5 (1-4)**	*0*	*0*	*0*
Range	*0-5*	*0-4*	*0-8*	*0-2*
Number of colonies tested	*20*	*20*	*20*	*20*

** 0.95 confidence limits of median.*

P<0.005, WSRT).

The molluscan corallivore *Jenneria* and *T. corallina* were also tested for resident crustacean responses through pairwise introductions (table 13). Resident *Trapezia* spp. and *A. lottini* again showed relatively low responses to foreign *Trapezia* (2.2 and 0.3 responses per 3 min. respectively), and equally low responses to *Jenneria* (P>0.05, for both comparisons, WSRT).

In addition to the marked differences in the frequency of aggression toward *Acanthaster* and crustacean competitors, some of the responses of resident *Trapezia* spp. toward corallivores and competitors were also qualitatively different. However, no different kinds of aggressive acts were detected in *Alpheus*.

A more detailed analysis of the types of agonistic responses elicited in resident *Trapezia* (*T. davaoensis*) by introduced (LAR) crustacean symbionts at Guam (table 11) is presented in fig. 7. These observations were probably biased toward more frequent high-level aggression because the tethered crustaceans were not entirely free to flee and thus avoid escalated fighting (Preston, 1973). (The *T. davaoensis* - *T. davaoensis* responses are not shown separately because only seven interactions were observed. These were nearly equally divided among the startle display, touching, and high-level aggressive responses, and are included in the mean bar graph plot.) High-level aggression (jerking, pinching, cutting of carapace or appendages, etc.) comprised about 30% of the responses. The startle display (lateral merus display of Wright, 1968), touching, and striking also represent frequent responses. Striking and pushing, usually accomplished with closed chelae, were not observed in encounters with *Acanthaster*. When coral guards contacted *Acanthaster* they invariably responded with a vigorous aggressive attack similar to the "wild fight" aggression described by Schöne (1968). Encounters among the crustacean symbionts often resulted in fighting, but usually this was more "formalized" or non-injurious than encounters with *Acanthaster*.

The frequency of all behavioral responses observed in *Trapezia* spp. toward introduced (LAR) *T. ferruginea* in Panama are shown in fig. 8. Because of the differing protocols employed in Guam and Panama, a comparison of the results shown in figs. 7 and 8 is not possible. (It is likely that chemical cues from *Acanthaster* aroused the crustaceans and elicited more vigorous responses.) The Panama results,

---

FIGURE 7. Proportion of various agonistic responses elicited from Trapezia *by introducing foreign shrimps and crabs into resident crab's coral host (Guam, see table 11).*

RESPONSES

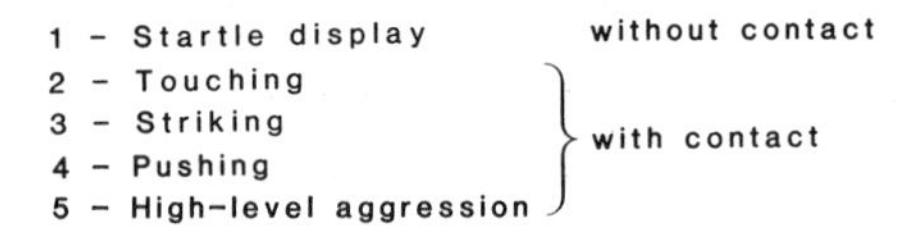
1 - Startle display
2 - Touching
3 - Striking
4 - Pushing
5 - High-level aggression
without contact
with contact

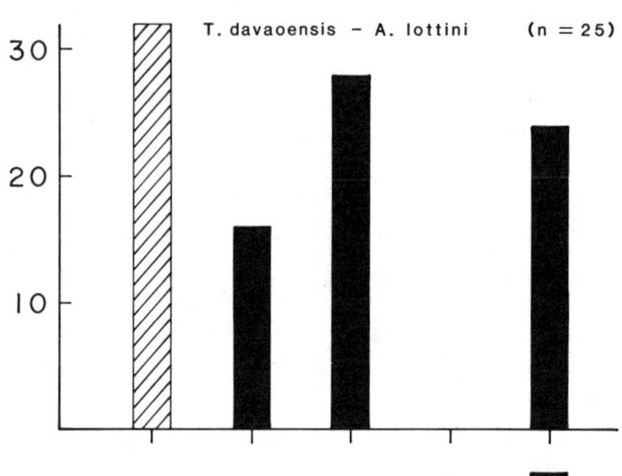
T. davaoensis - A. lottini (n = 25)
30
20
10

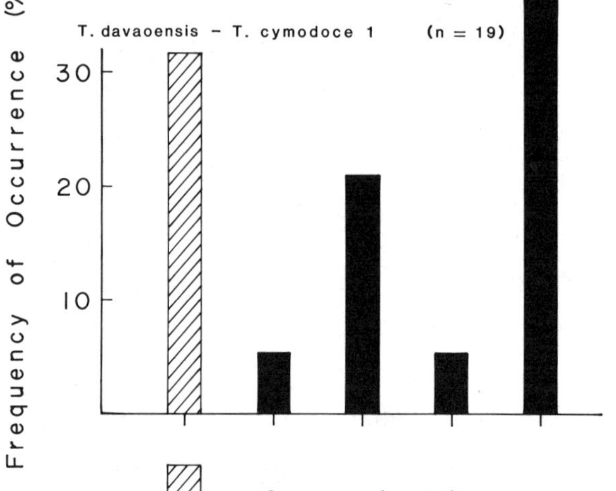
Frequency of Occurrence (%)
T. davaoensis - T. cymodoce 1 (n = 19)
30
20
10

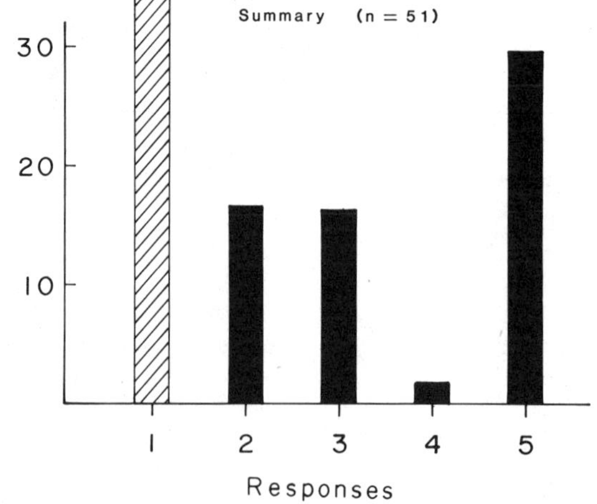
Summary (n = 51)
30
20
10
1
2
3
4
5
Responses

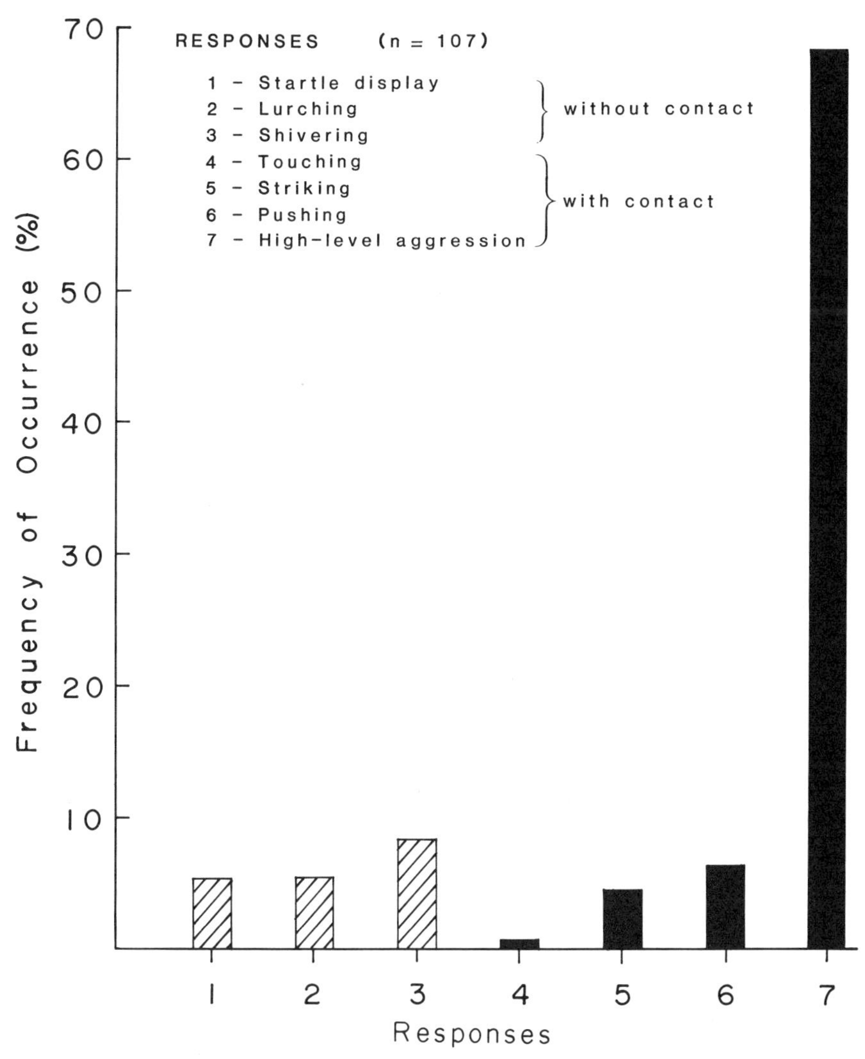

FIGURE 8. Proportion of various agonistic responses elicited from Trapezia *by introducing foreign crabs into resident crab's coral host (Panama, see table 12).*

however, show six different responses for resident crabs, in addition to high-level aggression. Lurching and shivering (rapid quivering of the entire body), responses not observed at Guam, occurred in 14% (n = 107) of the observations in Panama. Shivering has not been observed in encounters with sea star corallivores. Shivering movements have been reported previously in two other animals: a coral-inhabiting goby (Lassig, 1977) and a deep-sea galatheid (an anomuran crustacean; J. Parzefall in Weygoldt, 1977). The goby apparently signals its resident status by shivering, and the trembling cheliped movements in the galatheid may be involved in a courtship display. Finally, it is worth noting that none of the low-level agonistic responses (fig. 8, 1 through 6) was executed by *Trapezia* spp. toward *Acanthaster* (arm tip) during these observations. High-level aggression toward the corallivore was the rule.

It is reasonable to assume that these various displays are involved in several different functions. One could argue that high-level aggression is associated solely with crustacean social and competitive functions. I believe, however, that the high-level aggressive responses ("wild fighting"), which often cause physical damage, are employed mainly to thwart predators capable of killing the host coral. Low-level displays ("formalized fighting"; see Preston, 1973) and occasional high-level aggression (Abele, in press, and this study) seem sufficient to account for intraphyletic competitive interactions. Therefore, it is suggestive that the highest levels of aggression exercised by coral guards represent an evolved protective response induced by potent corallivores.

Benefits of a Coral Host

In addition to the support and shelter from predators provided by the coral skeleton, the nematocysts and the ciliary-mucoid/feeding-cleansing mechanisms of corals may protect crustacean symbionts from predators, parasites, and fouling organisms (Patton, 1976). Coral mucus, and its bacterial flora and entrapped detritus, is the chief food source of crustacean symbionts (Knudsen, 1967; Patton, 1974, 1976; Castro, 1976). High-energy lipids (wax esters), amino acids, proteins, mono- and polysaccharides have been identified in mucus (Benson and Muscatine, 1974; Ducklow and Mitchell, 1979). Purified mucus from a mushroom coral (after the removal of particulate debris and mineral salts) consisted of low fractions of lipid and protein of low caloric values, suggesting that extraneous, entrapped materials can add significantly to the nutrient value of mucus (Krupp, in press).

In terms of a possible effect of the crustacean symbionts on mucus production, it would be of interest to compare the quantity and quality of available mucus in a variety of host and non-host corals. The observation that in Panama crustacean symbionts, when offered a choice of hosts, do not associate with an abundant branching coral (*Psammocora stellata*), but always select *Pocillopora* spp. for habitation, suggests that the quantity and/or quality of mucus from the usual host are preferred. Of course, crustacean symbionts may also avoid nonpocilloporid corals because of a potent nematocyst defense and/or repellent secondary metabolites.

Pocilloporid and acroporid corals inhabited by crustacean symbionts produce large amounts of mucus. Apparently the "food brushes" and "food combs" on the walking legs of *Trapezia* and *Tetralia* (less well developed in this genus) can be used to stimulate mucus release and collection (Knudsen, 1967; Patton, 1974; Castro, 1976). The quantities of mucus collected from four coral species in Panama are summarized in table 14 and indicate that the host coral (*Pocillopora*) releases by far the largest crude volumes of mucus (unseparated mucus-seawater exudate). This result, in Panama, is in line

Table 14. Crude mucus-seawater exudate (ml/100 cm^2 coral surface/15 min.) collected from four coral species in Panama.

Species	*Colony form*	*Median*	*0.95 conf. lim. of median*	*Range*	*N*
Pocillopora damicornis	*Ramose*	*0.21*	*0.10-0.33*	*0.10-0.41*	*13*
Psammocora stellata *Verrill*	*Ramose*	*<0.10*	*<0.10-0.10*	*<0.10-0.17*	*12*
Pavona gigantea *Verrill*	*Massive*	*<0.10*	*<0.10-<0.10*	*<0.10-0.15*	*15*
Pavona varians *Verrill*	*Massive*	*<0.10*	*<0.10-<0.10*	*<0.10-<0.10*	*15*

with the high nectar volumes produced by (a) plants hosting ant guards (Janzen, 1966a) and (b) plants pollinated by birds (Stiles, 1980). The observed differences in mucus production do not agree, however, with the Red Sea results of Richman et al. (1975), who found that massive rather than branching corals generally released higher masses of mucus. No mention was made in their study of the presence of crustacean symbionts. Any relation that may exist between crustacean symbionts and mucus production will be complicated because mucus has at least two important demonstrated functions in corals, namely, ciliary-mucoid feeding (Goreau et al., 1971; Lewis and Price, 1975; Lewis, 1976, 1977), and cleansing activities (Hubbard and Pocock, 1972; Muscatine, 1973). Evidence supporting the coevolution of extrafloral nectaries in conjunction with the defensive behavior of ant guards is impressive, but not conclusive (Bentley, 1977b). A number of physiological functions have been suggested for nectar production, but none has been adequately substantiated.

A potentially important alternative food source for crustacean symbionts is presently under study by Stimson (pers.

comm.). He has identified "fat bodies" (lipid reserves) produced by two species of *Pocillopora*, but not by four coral species without crustacean symbionts. These fat bodies are released into the coelenteron and eventually become incorporated into mucus (possibly the reason for the high concentrations of a wax ester and triglycerides in *Pocillopora*; Benson and Muscatine, 1974). The significance of these fat bodies as a food source and their occurrence among protected and unprotected corals are of vital interest.

DISCUSSION

Ecological, behavioral, and evolutionary aspects of the crustacean-coral and ant-plant mutualisms are compared in order to develop a framework for inferring certain evolutionary developments in host coral defense.

Comparison of Guarded Coral and Myrmecophyte Mutualisms

There are numerous parallels, and some notable differences, between guarded corals and acacias. The swollen-thorn acacia/*Pseudomyrmex* mutualisms of the New World tropics and subtropics have been extensively studied and regarded as a highly coevolved interdependent system. Information on these particular ant-plant mutualisms is largely from Janzen (1966a and b, 1974); Bentley (1977a) provided an important source for additional examples of insect-mediated plant defenses.

First of all, it may be noted that several species are involved in the marine and terrestrial host-guarded mutualisms. In the present study, which did not attempt a comprehensive survey of all the species engaged in host-coral defense, 10 pocilloporid and 3 acroporid protected-coral species were observed. Although the number of valid species belonging to the Pocilloporidae is still unsettled (Wells, 1972; Veron and Pichon, 1976), a conservative estimate indicates the existence of 20 to 25 pocilloporid species that contain crustacean symbionts capable of host defense. Wallace (1978) estimates that around 70 species of *Acropora* will be recognized. Many of these species, perhaps at least one-half, will probably be shown to harbor *Tetralia* crabs. This estimate is based on the numerous ramose species in *Acropora* and the generally frequent association of *Tetralia* with branching *Acropora* colonies (Patton, 1966; Eldredge and Kropp, in press; pers. obs.). *Trapezia* contains at least 23 species (Serène, 1969), and all 14 species thus far examined engage in coral defense. Two species of *Tetralia* have been named and likely at least seven species will be recognized (Patton, 1966), based on color and fidelity in sexual pairing (see Castro, in press). Two color forms of *Tetralia glaberrima* show defensive behavior. There are more than 200 species of *Alpheus*, but only a minority is symbiotic (Bruce, 1976). Four *Alpheus* species live symbiotically with *Pocillopora damicornis* in Panama (Abele and Patton, 1976), and 10 with *P. damicornis* on the Great Barrier Reef (Austin et al., 1980). Perhaps as many as 20 species of *Alpheus* are obligate symbionts. However, only two species of *Alpheus* are known to defend their hosts from predators: *A. lottini* defends corals and *A. armatus* defends a sea anemone (Smith, 1977).

Although worldwide the legume genus *Acacia* comprises at least 750 species, in the Americas only 12 species belong to the swollen-thorn acacia group that harbors ants. Of the 150-200 species of New World *Pseudomyrmex*, at least 13 are obligate acacia ants capable of host defense. Many other species of plants and ants are loosely associated, with occasional to frequent visitation to foliar nectaries and at least some degree of plant protection (e.g., Bentley, 1976, 1977a,b).

Approximately 60 corals are defended by 16 obligate crustaceans, and 12 acacias are defended by 13 obligate ants. In terms of the number of potential and realized defended hosts, approximately 22% of the species belonging to two coral families (60 of 275 species) contain obligate crustacean symbionts, and approximately 2% of the acacia species (12 of 750 species) contain obligate ant symbionts. Probably the majority, if not all, of the *Trapezia* and *Tetralia* crabs are obligate symbionts on coral and engage at some level in host defense. The proportions of crustacean and ant symbionts that are confined to their hosts and demonstrate host defense differ by about an order of magnitude: 64% of the crustaceans (32 of 50 species) and 6% of the ants (13 of 200 species).

Swollen-thorn acacia ants are dependent on their hosts for food, as completely as crustacean symbionts are on their host corals. Deprived of their hosts, ants die from starvation and crustaceans either succumb to predation or starve.

Janzen (1966a) maintained that swollen-thorn acacias are dependent upon their obligate ant guards for survival. Unprotected acacias (ants removed) suffered higher mortality rates than protected acacias. Unprotected corals (crustaceans removed) are also subject to higher rates of predation than protected corals. However, I have observed numerous instances of predation on protected corals, especially during severe *Acanthaster* outbreaks. Many protected corals are not defended for short periods (table 6), and numerous protected corals are overwhelmed (eaten whole or in part) by *Acanthaster* foraging in large swarms (100 or more individuals per hectare). Under such conditions, selection for high-level aggression among the crustacean guards must be intense. Moreover, such predation pressure would appear to satisfy the level of host survival (intermediate survivorship) postulated by Roughgarden (1975) for the evolution of mutualisms. It should also be noted that not all adult pocilloporid corals harbor aggressive symbionts. For example, Preston (1973) reported 22% of *Pocillopora meandrina* lacking any *Trapezia* in Hawaii, and Stimson (pers. comm.) found a population of *P. damicornis* in Hawaii with most colonies lacking *Trapezia*. These observations suggest that swollen-thorn acacias might depend more on their ants for survival than protected corals do on their crustaceans. It is possible that ant plants are subject to more intense predation and interference competition than protected corals (see below).

Some further similarities related to host protection are significant. Host protection in both corals and plants is effective during the day and night, because predation in marine and terrestrial environments occurs at all times of day. Ants detect intruders visually during the day and by contact at night. Chemical and, to a lesser extent, visual cues elicit defensive behavior in crustaceans (Glynn, 1980). In some ant-plant mutualisms, continuous host protection is provided by ants active at different times of the day (Bentley, 1977a). *Pseudomyrmex* is constantly active and alert. All species of crab and shrimp guards are alert at all hours and quickly assume the defensive mode whenever approached by *Acanthaster*.

Corallivores or herbivores attacking a coral or plant host are often damaged by the defending symbionts; but seldom, if ever, are the predators or their parts eaten. The symbionts rely exclusively on the products of their hosts for sustenance. Only rarely are pieces from dismembered insect folivores used as food by ant larvae.

In spite of the presence of crustaceans and ants, the coral and plant hosts are preyed upon by certain species that are able to circumvent the symbionts' defenses. Insect herbivores overcome ant defenses by (1) an impenetrable cuticle and no avoidance reaction to ants, (2) being totally ignored by the ants, and (3) being able to throw off the attacking ants. Corallivores that feed on defended *Pocillopora* do so chiefly by means of 1 and/or 2 above. It is largely unknown which of these traits are the principal deterrents to crustacean defense. Although many corallivores possess heavy armature, exoskeletons or shells, e.g., *Eucidaris* (sea urchin), *Culcita* (cushion star), *Trizopagurus* (hermit crab), and *Jenneria* (snail), they still normally expose soft parts and delicate organs such as tube feet, respiratory structures, stomach, antennae, eyes, etc. All of these structures are easily damaged by crustaceans engaged in high-level defense. Some corallivores are essentially ignored, e.g., *Eucidaris*, which feeds on *Pocillopora* spp. in the Galapagos Islands (Glynn and Wellington, in press), and pufferfishes (*Arothron* spp.). *Acanthaster* and *Culcita* that successfully feed on protected corals often do so by losing spines, tube feet, and papulae, and occasionally by receiving lacerations on the extruded stomach and body wall. Crustacean coral guards may temporarily, for unknown reasons, suspend their aggressive defensive behavior, but I have never seen the guards repulsed by a corallivore.

Ant and crustacean guards tend to ignore certain predators that can cause damage to the host. To what extent such avoidance responses relate to (1) the importance of predators (i.e., to their efficiency in causing host mortality), (2) the recency of appearance of predators in evolutionary time (as population outbreaks, foreign invaders or new adaptive types), or (3) the deterrent properties of predators (e.g., mechanical and chemical defenses), are all possibilities in need of study. *Jenneria*, an effective endemic predator of *Pocillopora* in the eastern Pacific, is virtually ignored by crustacean guards. Is this because (a) the corallivore is unimportant as a selective agent, (b) the corallivore discourages crustacean attacks through mucus secretions, (c) the corallivore is impervious, (d) the crustacean guards or their hosts are relatively recent arrivals from the central Pacific, or (e) is it due to some other cause(s)?

Protected corals and plants are readily eaten when their guards are removed. Both hosts assume higher positions in their respective prey-preference hierarchies of consumers when deprived of guards. Janzen (1966a) interpreted this to imply that certain mechanical and chemical defenses characteristic of acacias (present in acacias without obligate ant guards) have been lost by swollen-thorn acacias whose main line of defense against herbivores is assumed by ant guards. This will be difficult to investigate among the Pocilloporidae because all reef-building species in this coral family (in the Indo-Pacific region) seem to possess crustacean guards. I doubt that the mutualism, where present among *Acropora* species, is sufficiently evolved in the Acroporidae to reveal any differences between corals with and without guards.

One notable difference between the coral and acacia mutualisms is the presence of specialized morphological structures, presumably evolved by acacias to benefit their ant symbionts. These structures are the swollen thorns (used as domatia), enlarged foliar glands (that produce nectar), and Beltian bodies (that produce nutrients of a proteinaceous and fatty nature). The presence on corals of specialized structures that solely benefit crustacean symbionts has not been demonstrated. There are, however, possible enhancements (or reductions) of existing structures or functions that might better serve a coral's crustaceans, e.g., tightly packed branches, which provide numerous small spaces that can be utilized by recruiting juvenile and subadult crustaceans, abundant supply of high quality mucus, fat body synthesis and release, and reduced potency of nematocysts and chemical defenses. It is also possible that the existing design (preadaptations) of corals is sufficient to support crustacean symbionts and new traits are not needed.

As for the guest symbionts, the degree of their commitment to the respective mutualisms appears to be approximately equal. Many of the preadaptations of crustaceans and ants that enhance the development of host protection, such as agility, vision, possession of offensive structures (strong mandibles, stings, chelipeds, snapping claws), etc., are shared by both arthropod groups. Certain specialized structures present in the crustacean guards, but not unique to them, may represent modifications which provide for more efficient coral habitation. The crabs' reinforced dactyls (Borradaile in Patton, 1976), permit a tenacious grip on the coral (fig. 2E), and, on their legs, food brushes and combs collect mucus (Knudsen, 1967). The feeding behavior of *Trapezia* guards, insertion of legs into coral polyps to stimulate mucus flow (Knudsen, 1967), has not been reported in other crabs. [However, two species of symbiotic hapalocarcinid crabs have been observed recently to prod their pocilloporid hosts (R.K. Kropp, pers. comm.) much the same as reported by Knudsen (1967).] The marked aggressive behavior of decapod crustaceans in general (Schöne, 1968; Dingle and Caldwell, 1969; Bruce,1976) is well suited to their role in host defense.

Ants differ markedly from crustaceans in their aggressiveness toward encroaching vegetation. Obligate acacia ants bite and/or sting any foreign vegetation that comes into contact with their host plant. My observations indicate that crustacean guards ignore all vegetation adjacent to, or actually growing directly on, the live branches of host colonies. Study of coral (*Pocillopora damicornis*) survivorship, growth, and comparisons between colonies with or without (removed) crustacean symbionts, indicate no significant differences in algal biomass accumulation on either the live or dead branches of corals. Algal spores are probably efficiently removed from live branches by the self-cleansing activities of corals. This, and the rarity of algal encroachment on corals, suggests that crustacean guards ignore plants because they seldom pose a threat to their hosts. Perhaps the intense activities of herbivores on coral reefs, and their effectiveness in limiting algal growth (e.g., Stephenson and Searles, 1960; Bakus, 1969; Sammarco et al., 1974; Vine, 1974; Wanders, 1977; Brock, 1979), are the ultimate causes of the crustacean's unresponsiveness toward vegetation.

Whereas the entire life cycle of obligate ants is passed on acacias, eggs carried by female crabs and shrimp hatch into larvae which spend some time, perhaps several weeks (Garth, 1974), in the plankton. It is unknown whether metamorphosing larvae settle on live coral or on other substrates and later migrate to coral. Abele (in press) found that in the western Pacific juvenile and small adult crustaceans were the usual colonists.

Two additional differences in the host defense mutualisms are notable. Excited ant workers liberate an alarm odor (pheromone) which rapidly draws ants to the intruder. Crustacean guards respond primarily to the secretions of attacking *Acanthaster*, at least initially. Colony cohorts probably perceive the activities of individuals engaged in defense, and thus may be partly aroused by visual and acoustic cues. However, I could not find any trend for a reinforcement effect between defending crabs and shrimps.

In a group like ants with a reproductive caste, it is not surprising that non-reproductive workers are in the vanguard of host defense. In crustacean symbionts, gravid crabs and shrimps are often among the first individuals to confront and attack intruders. Female crab and shrimp guards are slightly larger than males, 9% (n = 47 pairs) and 6% (n = 22) respectively (Patton, 1974), and therefore have a size advantage in defense. Moreover, the frequency of host defense demonstrated by female coral guards could also be a result of the high-level aggression often associated with individuals with a high reproductive effort. Defending females would not be directly at risk with most corallivores (e.g., sea stars and snails).

It is now possible to consider more explicitly the question of whether corals and crustacean guards occur together by chance or whether their mutually beneficial traits have resulted from concurrent and interdependent evolutionary change (coevolution).

The coral host does not show many traits that suggest a direct and unique accommodation to the crustacean symbionts. Corals provide shelter and mucus food, but there is little indication that these resources are superior in coral hosts as compared with ramose corals not harboring crustacean guards. Since the crustaceans are highly host selective, not inhabiting other seemingly suitable corals, it is possible that their usual coral hosts are offering certain special advantages. That such attributes are coevolved is certainly plausible, but this assumption is in urgent need of careful testing. The high palatability of host corals without their guards suggests a diminution in structural and chemical defenses in conjunction with the possession of defending crustaceans. Fat body production by host corals is possibly a unique development and deserves further study in this connection.

Evidence for traits evolved for symbiosis among the crustaceans appears more convincing. The feeding behavior and modified appendages for mucus collection represent a distinctive combination of features for dealing with this specialized food resource. Perhaps of greatest significance is the high-level defensive behavior of the crustacean guards, which is directed most strongly (in frequency and intensity) toward the most potent corallivores. Thus, evidence of coevolution is most apparent in the behavioral traits of the crustacean guards, in contrast to the ant-acacia system which shows greatest modifications in the structural attributes of the host plants.

Perhaps an important result of this study is the recognition that host-guard mutualisms, and particularly the crustacean-coral example, is apparently subject to numerous selective forces. Some of the problematical results of this study -- for example, strong host defense under low predation pressure (Pearl Islands), and weak host defense toward an important specialist corallivore (*Jenneria*) -- may be due in part to the many ecological and behavioral pressures involved. The variety of corallivores, their methods of attack, and effects on local coral populations at any given time, are enormously diverse. The evolution of anti-predatory defenses, within particular coral species, must also be a vital component of the system. Fish and octopod predators are abundant: resident crustaceans must protect their coral host -- being sufficiently forceful to repel corallivores, but also exercising extreme caution to avoid predatory attacks on themselves. Competitive interactions represent another important element in the mutualism. Shelter sites and mucus food supplied by the coral should be protected from innumerable competing species. In the extreme, competition blends imperceptibly with predation, e.g., when a corallivore kills the host coral. And then there are numerous complex social interactions (e.g., territorial and sexual behaviors) that have not been examined in this study. The analysis of particular coevolved systems, and the recognition of presumed coevolved traits, has probably been greatly complicated by a multitude of effects operating at the community level (Fox, 1981; Howe, 1980).

Some Speculations on the Evolution of Crustacean Defense

Several associations are suggestive of certain phases through which the coral-guard mutualism may have passed in its evolution. Numerous crustaceans are facultative coral associates that seek shelter in branching corals but feed on other organisms on or beyond the coral colony. Abele (in press) estimated that at mainland eastern Pacific localities about two-thirds of the decapod crustacean species found on the live branches of pocilloporid corals also occur in other habitats. [The proportion of facultative pocilloporid associates is apparently relatively low in oceanic environments, e.g., 21% on an atoll, but still considerable (Kropp and Birkeland, in press).] I have observed facultative crustaceans defending their coral hosts in a seemingly haphazard manner. For example, a stomatopod species (*Gonodactylus zacae* Manning) inhabiting the coral *Psammocora stellata* was observed to strike (with its raptorial appendage) and repel an *Acanthaster* attempting to mount the colony. I have also observed repeatedly the xanthid crab *Heteractaea lunata* (H. Milne Edwards and Lucas) to strike and push the gastropod corallivore *Jenneria* from its coral host (*Pocillopora damicornis*). The striking and pushing acts, executed with closed chelipeds, were similar to those observed in *Trapezia* (fig. 2B). *Heteractaea* feeds chiefly on crustaceans associated with corals and, in lesser amounts, on detritus and coral mucus.

Coral mucus is a highly nutritive food, but the relative scarcity of species feeding on it suggests that it may be difficult to process. When removed from the coral, mucus invariably contains numerous stinging capsules (nematocysts).

Nematocysts are always abundant in the stomachs of mucus-feeding species (Castro, 1976; Reese, 1977). Therefore, it seems that the ability to utilize mucus as a food may have been an important achievement in the evolution of obligate coral symbionts. If, as a consequence of or incidental to mucus feeding (e.g., through more efficient cleansing activities or corallivore deterrence), the consumer species increased the fitness of the coral then the association will become reciprocal and mutually beneficial. Instead of an escalation in the physical and chemical defenses of the food-providing species, and counteradaptations in the consumer species (as noted among predator-prey associations in insects and plants; Ehrlich and Raven, 1964; Feeny, 1975), there could evolve a highly specialized and interdependent mutualism. This is a plausible explanation for the weak defenses of swollen-thorn acacias divested of their ant guards (Janzen, 1966a), and may also account for the relatively high palatability of pocilloporid corals (Glynn et al., 1972; Glynn, 1982; Reese, 1977).

Many, and perhaps the majority, of coral mucus feeders do not protect corals from corallivores. How coral defense may have been initiated in obligate xanthid crabs can be seen in *Tetralia*. *Tetralia* crabs are relatively small and possess only moderately developed chelae. At least one species of *Tetralia* is capable of weak host defense. *Tetralia*'s defensive behavior is somewhat erratic, but when executed is sufficient to deter at least some corallivores. The evolution of effective coral guards (*Trapezia* spp. and *Alpheus lottini*) could have been induced by species competing for shelter and food, and/or by corallivores that killed part or all of the host colony. The crustacean-coral mutualism is perhaps the result of reciprocal evolutionary responses in the manner proposed by Janzen (1966a) for the ant-acacia interaction system: mucus production would support crustaceans which would benefit the coral (protection from predators), and the coral would produce more and possibly higher quality mucus; eventually the crustaceans would develop a trophic dependency on the coral host. The evolution of a potent defensive behavior was probably an important development in this system, not only for protection against corallivores, but also to deter competing mucus feeders.

I conclude with the question of whether coral guards have influenced the diversification of the Acroporidae and Pocilloporidae. The idea that the appearance of animal pollinators and dispersers was causally related to the radiation of flowering plants has received wide attention (e.g., Snow, 1980, 1981). I am inclined to the view that protected coral mutualisms have had no appreciable effect on coral species diversification.

Together with the Poritidae (suborder Fungiina), the Acroporidae and Pocilloporidae account for more than two-thirds of contemporary reef-building corals (Wells, 1956). These groups expanded greatly during the Late Tertiary. No crustacean coral guards are known in extant members of the Poritidae, although many taxa in this family have ramose colonies that could provide shelter. *Montipora* is a diverse genus in the Acroporidae; at Guam it contains at least 30 species, compared with 37 species of *Acropora* (see Randall, 1981). Some species of *Montipora* have ramose and foliaceous colonies, yet no crustacean guards are associated with these corals. The great success of *Acropora* during Pleistocene and Recent times is also difficult to reconcile with the weak (at best) crab defense observed in species in this group. Considering the significant effect of potent crustacean guards on the survivorship of pocilloporid corals, it is conceivable that this level of defense, once attained, would favor the persistence and diversification of such mutualisms over evolutionary time.

ACKNOWLEDGEMENTS

I am grateful to the Honorable Peter T. Coleman, Governor of American Samoa, and to Henry Sesepasara, Director, Office of Marine Resources, for the opportunity to work in Samoa. The following kindly assisted with logistic support and field work in Samoa: Patrick G. Bryan, William Pedro, Roger Pflum, Richard C. Wass, and Gerard M. Wellington. I am indebted to Charles Birkeland, Director, Marine Laboratory, University of Guam, for providing the opportunity to work on Guam. Especially helpful at Guam were Mitchell W. Colgan, Roy K. Kropp, Henry Moore, Olga M. Odinetz, Gyongyi Plucer Rosario, and Vaughan Tyndzik. I thank Ira Rubinoff, Director, Smithsonian Tropical Research Institute, for encouragement and support in Panama, and also Raymond Highsmith and Anibal Velarde for help in the field. Coral identifications in Panama, and one species from Samoa, are due to J.W. Wells; and in Samoa and Guam, to R.H. Randall. P. Castro identified the crustacean symbionts from Panama; J.S. Garth and R.B. Manning, those from Samoa; and O.M. Odinetz and R.K. Kropp, those from Guam. R.B. Manning also identified a stomatopod from Panama. I also thank Arcadio Rodaniche for help in improving the drawings in fig. 2. This paper has benefited greatly from discussions with Harilaos Lessios, Sally C. Levings, Martin H. Moynihan, Michael H. Robinson, and Jane A. Sherfy, and from the critical comments offered by R.K. Kropp and G.J. Vermeij. The Scholarly Studies Program, Smithsonian Institution, provided financial assistance in Guam and Panama.

LITERATURE CITED

ABELE, L.G. 1976. Comparative species richness in fluctuating and constant environments: Coral-associated decapod crustaceans. Science, 192:461-463.

ABELE, L.G. In press. Biogeography, colonization and experimental community structure of coral-associated crustaceans. *In:* Strong, D., L.G. Abele, and D. Simberloff, eds., Ecological Communities: Conceptual Issues and the Evidence. Princeton University Press.

ABELE, L.G., and W.K. PATTON. 1976. The size of coral heads and the community biology of associated decapod crustaceans. Journal of Biogeography, 3:35-47.

AUSTIN, A.D., S.A. AUSTIN, and P.F. SALE. 1980. Community structure of the fauna associated with the coral *Pocillopora damicornis* (L.) on the Great Barrier Reef. Australian Journal of Marine and Freshwater Research, 31:163-174.

BAKUS, G.J. 1969. Feeding and energetics in shallow marine waters. International Review of General and Experimental Zoology, 4:275-369.

BENSON, A.A., and L. MUSCATINE. 1974. Wax in coral mucus: Energy transfer from corals to reef fishes. Limnology and Oceanography, 19:810-814.

BENTLEY, B.L. 1976. Plants bearing extrafloral nectaries and the associated ant community: Interhabitat differences in the reduction of herbivore damage. Ecology, 57:815-820.

BENTLEY, B.L. 1977a. The protective function of ants visiting the extrafloral nectaries of *Bixa orellana* (Bixaceae). Journal of Ecology, 65:27-38.

BENTLEY, B.L. 1977b. Extrafloral nectaries and protection by pugnacious bodyguards. Annual Review of Ecology and Systematics, 8:407-427.

BIRKELAND, C., and R.H. RANDALL. 1979. Report on the *Acanthaster planci* (*Alamea*) studies on Tutuila, American Samoa. Report submitted to Office of Marine Resources, Pago Pago, American Samoa.

BIRKELAND, C. In press. Terrestrial runoff as a cause of outbreaks of *Acanthaster planci* (Echinodermata: Asteroidea). Marine Biology (Berlin).

BROCK, R.E. 1979. An experimental study on the effects of grazing by parrotfishes and role of refuges in benthic community structure. Marine Biology (Berlin), 51:381-388.

BRUCE, A.J. 1976. Shrimps and prawns of coral reefs, with special reference to commensalism; pp. 37-94. *In*: Jones, O.A., and R. Endean, eds., Biology and Geology of Coral Reefs. Vol. 3, Biology 2, New York: Academic Press.

CASTRO, P. 1976. Brachyuran crabs symbiotic with scleractinian corals: A review of their biology. Micronesica, 12:

99-110.
CASTRO, P. In press. Notes on symbiotic decapod crustaceans from Gorgona Island, Colombia, with a preliminary revision of the eastern Pacific species of *Trapezia* (Brachyura, Xanthidae), symbionts of scleractinian corals. Anales del Instituto de Investigaciones Marinas de Punta de Betin, 12.
CHESHER, R.H. 1969. Destruction of Pacific corals by the sea star *Acanthaster planci*. Science, 165:280-283.
CHESHER, R.H. 1972. The status of knowledge of Panamanian echinoids, 1971, with comments on other echinoderms; pp. 139-158. *In*: Jones, M.L., ed., The Panamic Biota: Some Observations Prior to a Sea-level Canal. Bulletin of the Biological Society of Washington, no. 2.
CONNELL, J.H. 1973. Population ecology of reef-building corals; pp. 205-245. *In*: Jones, O.A., and R. Endean, eds., Biology and Geology of Coral Reefs, Vol. 2, Biology 1, New York: Academic Press.
DANA, T., and A. WOLFSON. 1970. Eastern Pacific crown-of-thorns starfish populations in the lower Gulf of California. Transactions of the San Diego Society of Natural History, 16(4):83-90.
DANIEL, W.W. 1978. Applied nonparametric statistics. Boston: Houghton Mifflin Co. 510 pp.
DINGLE, H., and R.L. CALDWELL. 1969. The aggressive and territorial behavior of the mantis shrimp *Gonodactylus bredini* Manning (Crustacea: Stomatopoda). Behavior, 33: 115-136.
DUCKLOW, H.W., and R. MITCHELL. 1979. Composition of mucus released by coral reef coelenterates. Limnology and Oceanography, 24:706-714.
DUNN, D.F. 1981. The clownfish sea anemones: Stichodactylidae (Coelenterata: Actiniaria) and other sea anemones symbiotic with pomacentrid fishes. Transactions of the American Philosophical Society, 71:1-113.
EHRLICH, P.R., and P.H. RAVEN. 1964. Butterflies and plants: A study in co-evolution. Evolution, 18:586-608.
ELDREDGE, L.G., and R.K. KROPP. In press. Crustacean-induced skeletal modification in *Acropora* with reference to an undescribed Indo-West Pacific crab species. Proceedings of the Fourth International Coral Reef Symposium, Manila.
ENDEAN, R. 1973. Population explosion of *Acanthaster planci* and associated destruction of hermatypic corals in the Indo-West Pacific region; pp. 389-438. *In*: Jones, O.A., and R. Endean, eds., Biology and Geology of Coral Reefs, Vol. 2, Biology 1, New York: Academic Press.
ENDEAN, R., and R.H. CHESHER. 1973. Temporal and spatial distribution of *Acanthaster planci* population explosions in the Indo-West Pacific region. Biological Conservation, 5(2):87-95.

FEENY, P. 1975. Biochemical coevolution between plants and their insect herbivores; pp. 3-19. *In*: Gilbert, L.E., and P.H. Raven, eds., Coevolution of Animals and Plants. Austin: University of Texas Press.

FORSBERGH, E.D. 1969. On the climatology, oceanography and fisheries of the Panama Bight. Inter-American Tropical Tuna Commission Bulletin, 14(2):49-385.

FOX, L.R. 1981. Defense and dynamics in plant-herbivore systems. American Zoologist, 21:853-864.

GARTH, J.S. 1974. On the occurrence in the eastern tropical Pacific of Indo-West Pacific decapod crustaceans commensal with reef-building corals. Proceedings of the Second International Coral Reef Symposium, Brisbane, 1:397-404.

GLYNN, P.W. 1974. The impact of *Acanthaster* on corals and coral reefs in the eastern Pacific. Environmental Conservation, 1:237-246.

GLYNN, P.W. 1976. Some physical and biological determinants of coral community structure in the eastern Pacific. Ecological Monographs, 46:431-456.

GLYNN, P.W. 1980. Defense by symbiotic Crustacea of host corals elicited by chemical cues from predator. Oecologia (Berlin), 47:287-290.

GLYNN, P.W. 1982. Coral communities and their modifications relative to past and prospective Central American seaways. Advances in Marine Biology, 19:91-132.

GLYNN, P.W. In press. *Acanthaster* population regulation by a shrimp and a worm. Proceedings of the Fourth International Coral Reef Symposium, Manila.

GLYNN, P.W., and G.M. WELLINGTON. In press. Corals and coral reefs of the Galapagos Islands. Berkeley: University of California Press.

GLYNN, P.W., E.M. DRUFFEL, and R.B. DUNBAR. Manuscript. A dead Central American coral reef tract: Possible link with the Little Ice Age.

GLYNN, P.W., R.H. STEWART, and J.E. MCCOSKER. 1972. Pacific coral reefs of Panama: Structure, distribution and predators. Geologische Rundschau, 61:483-519.

GLYNN, P.W., H. VON PRAHL, and F. GUHL. In press. Coral reefs of Gorgona Island, Colombia, with special reference to corallivores and their influence on community structure and reef development. Anales del Instituto de Investigaciones Marinas de Punta de Betin, 12:

GLYNN, P.W., G.M. WELLINGTON, and C. BIRKELAND. 1979. Coral reef growth in the Galapagos: Limitation by sea urchins. Science, 203:47-49.

GOREAU, T.F., N.I. GOREAU, and C.M. YONGE. 1971. Reef corals: Autotrophs or heterotrophs? Biological Bulletin (Woods Hole), 141:247-260.

GOREAU, T.F., N.I. GOREAU, and C.M. YONGE. 1973. On the utilization of photosynthetic products from zooxanthellae and of a dissolved amino acid in *Tridacna maxima* f. *elongata* (Mollusca: Bivalvia). Journal of Zoology, 169: 417-454.

GOREAU, T.F., J.C. LANG, E.A. GRAHAM, and P.D. GOREAU. 1972. Structure and ecology of the Saipan reefs in relation to predation by *Acanthaster planci* (Linnaeus). Bulletin of Marine Science, 22:113-152.

HOWE, H.F. 1980. Monkey dispersal and waste of a neotropical fruit. Ecology, 61:944-959.

HUBBARD, J.A.E.B., and Y.P. POCOCK. 1972. Sediment rejection by recent scleractinian corals: A key to palaeo-environmental reconstruction. Geologische Rundschau, 61:598-626.

HUGHES, T.P., and J.B.C. JACKSON. 1980. Do corals lie about their age? Some demographic consequences of partial mortality, fission, and fusion. Science, 209:713-715.

JANZEN, D.H. 1966a. Coevolution of mutualism between ants and acacias in Central America. Evolution, 20(3):249-275.

JANZEN, D.H. 1966b. The interaction of the bull's-horn acacia (*Acacia cornigera* L.) with one of its ant inhabitants (*Pseudomyrmex ferruginea* F. Smith) in eastern Mexico. University of Kansas Science Bulletin, 47:315-558.

JANZEN, D.H. 1974. Swollen-thorn acacias of Central America. Smithsonian Contributions to Botany, 13:1-131.

JANZEN, D.H. 1980. When is it coevolution? Evolution, 34(3): 611-612.

KNUDSEN, J.W. 1967. *Trapezia* and *Tetralia* (Decapoda, Brachyura, Xanthidae) as obligate ectoparasites of pocilloporid and acroporid corals. Pacific Science, 21:50-57.

KROPP, R.K., and C. BIRKELAND. In press. Comparison of crustacean associates of *Pocillopora* from high islands and atolls. Proceedings of the Fourth International Coral Reef Symposium, Manila.

KRUPP, D.A. In press. The composition of mucus from the mushroom coral, *Fungia scutaria*. Proceedings of the Fourth International Coral Reef Symposium, Manila.

LASSIG, B.R. 1977. Communication and coexistence in a coral community. Marine Biology (Berlin), 42:85-92.

LEWIS, J.B. 1976. Experimental tests of suspension feeding in Atlantic reef corals. Marine Biology (Berlin), 36: 147-150.

LEWIS, J.B. 1977. Suspension feeding in Atlantic reef corals and the importance of suspended particulate matter as a food source; pp. 405-408. *In*: Taylor, D.L., ed., Proceedings of the Third International Coral Reef Symposium, Miami.

LEWIS, J.B., and W.S. PRICE. 1975. Feeding mechanisms and feeding strategies of Atlantic reef corals. Journal of

Zoology, 176:527-544.

LOSEY, G.S. 1972. Predation protection in the poison-fang blenny, *Meiacanthus atrodorsalis*, and its mimics, *Ecsenius bicolor* and *Runula laudandus* (Blenniidae). Pacific Science, 26:129-139.

MARISCAL, R.N. 1966. The symbiosis between tropical sea anemones and fishes: A review; pp. 157-171. *In*: Bowman, R.I., ed., The Galapagos. Berkeley and Los Angeles: University of California Press.

MAYOR, A.G. 1924. Growth-rate of Samoan corals. Papers from the Department of Marine Biology of the Carnegie Institution of Washington, 19:51-72.

MUSCATINE, L. 1973. Nutrition of corals. *In*: Jones, O.A., and R. Endean, eds., Biology and Geology of Coral Reefs, Vol. 2, Biology 1, New York: Academic Press.

MUSCATINE, L., and J.W. PORTER. 1977. Reef corals: Mutualistic symbioses adapted to nutrient-poor environments. BioScience, 27:454-460.

ORMOND, R.F.G., N.J. HANSCOMB, and D.H. BEACH. 1976. Food selection and learning in the crown-of-thorns starfish, *Acanthaster planci* (L.). Marine Behavior and Physiology, 4:93-105.

PATTON, W.K. 1966. Decapod Crustacea commensal with Queensland branching corals. Crustaceana, 10:271-295.

PATTON, W.K. 1974. Community structure among the animals inhabiting the coral *Pocillopora damicornis* at Heron Island, Australia; pp. 219-243. *In*: Vernberg, W.B., ed., Symbiosis in the Sea. Columbia, South Carolina: University of South Carolina Press.

PATTON, W.K. 1976. Animal associates of living corals; pp. 1-36. *In*: Jones, O.A., and R. Endean, eds., Biology and Geology of Coral Reefs. Vol. 3, Biology 2. New York: Academic Press.

PEARSON, R.G., and R. ENDEAN. 1969. A preliminary study of the coral predator *Acanthaster planci* (L.) (Asteroidea) on the Great Barrier Reef. Fisheries Notes (Department of Harbours and Marine, Queensland), 3:27-68.

PORTER, J.W. 1972. Predation by *Acanthaster* and its effect on coral species diversity. American Naturalist, 106: 487-492.

PORTER, J.W. 1974. Community structure of coral reefs on opposite sides of the Isthmus of Panama. Science, 186: 543-545.

POTTS, G.W. 1973. The ethology of *Labroides dimidiatus* (Cuv. and Val.) (Labridae: Pisces) on Aldabra. Animal Behaviour, 21:250-291.

PRESTON, E.M. 1973. A computer simulation of competition among five sympatric congeneric species of xanthid crabs. Ecology, 54(3):469-483.

RANDALL, R.H. 1973. Distribution of corals after *Acanthaster planci* (L.) infestation at Tanguisson Point, Guam. Micronesica, 9(2):213-222.

RANDALL, R.H. 1981. Preliminary checklist of the Scleractinia of Guam; pp. 26-34. *In*: A Working List of Marine Organisms from Guam. Technical Report Number 70, University of Guam Marine Laboratory.

REESE, E.S. 1977. Coevolution of corals and coral feeding fishes of the family Chaetodontidae. Proceedings of the Third International Coral Reef Symposium, Miami, Biology, 1:267-274.

RICHMAN, S., Y. LOYA, and L.B. SLOBODKIN. 1975. The rate of mucus production by corals and its assimilation by the coral reef copepod *Acartia negligens*. Limnology and Oceanography, 20:918-923.

ROBERTSON, R. 1970. Review of the predators and parasites of stony corals, with special reference to symbiotic prosobranch gastropods. Pacific Science, 24:43-54.

ROUGHGARDEN, J. 1975. Evolution of marine symbiosis--a simple cost-benefit model. Ecology, 56:1201-1208.

RUSSELL, B.C., G.R. ALLEN, and H.R. LUBBOCK. 1976. New cases of mimicry in marine fishes. Journal of Zoology, 180: 407-423.

SAMMARCO, P.W., J.S. LEVINTON, and J.C. OGDEN. 1974. Grazing and control of coral reef community structure by *Diadema antillarum* Philippi (Echinodermata: Echinoidea): A preliminary study. Journal of Marine Research, 32(1):47-53.

SCHÖNE, H. 1968. Agonistic and sexual display in aquatic and semi-terrestrial brachyuran crabs. American Zoologist, 8: 641-654.

SERÈNE, R. 1969. Observations on species of the group *Trapezia rufopunctata-maculata*, with a provisional key for all the species of *Trapezia*. Journal of the Marine Biological Association of India, 11:126-148.

SLOBODKIN, L.B., and L. FISHELSON. 1974. The effect of the cleaner-fish *Labroides dimidiatus* on the point diversity of fishes on the reef front at Eilat. American Naturalist, 108:369-376.

SMITH, W.L. 1977. Beneficial behavior of a symbiotic shrimp to its host anemone. Bulletin of Marine Science, 27(2): 343-346.

SNOW, D.W. 1980. Regional differences between tropical floras and the evolution of frugivory. Symposium on coevolutionary systems in birds (1978). Proceedings of the XVII International Ornithological Congress, Berlin, 1192-1198.

SNOW, D.W. 1981. Tropical frugivorous birds and their food plants: a world survey. Biotropica, 13(1):1-14.

STEPHENSON, W., and R.B. SEARLES. 1960. Experimental studies on the ecology of intertidal environments at Heron Island. Australian Journal of Marine and Freshwater Research, 11: 241-267.

STILES, F.G. 1980. Ecological and evolutionary aspects of bird-flower coadaptations. Symposium on co-evolutionary systems in birds (1978). Proceedings of the XVII International Ornithological Congress, Berlin, 1173-1178.

TAMURA, T., and Y. HADA. 1932. Growth rate of reef building corals, inhabiting in the South Sea Island. Science Reports of the Tohoku University, Fourth Series (Biology), 7: 433-455.

TRENCH, M.E., R.K. TRENCH, and L. MUSCATINE. 1970. Utilization of photosynthetic products of symbiotic chloroplasts in mucus synthesis by *Placobranchus ianthobupsus* (Gould), Opisthobranchia, Sacoglossa). Comparative Biochemistry and Physiology, 37:113-117.

TRENCH, R.K. 1969. Chloroplasts as functional endosymbionts in the mollusc *Tridachia crispata* (Bergh). Opisthobranchia. Sacoglossa. Nature, 222:1071-1072.

TRENCH, R.K., R.W. GREENE, and B.G. BYSTROM. 1969. Chloroplasts as functional organelles in animal tissues. Journal of Cell Biology, 42:404-417.

VERON, J.E.N., and M. PICHON. 1976. Scleractinia of eastern Australia. Part I. Families Thamnasteriidae, Astrocoeniidae, Pocilloporidae. Australian Institute of Marine Science, Monograph Series, 1:1-86.

VINE, P.J. 1970. Densities of *Acanthaster planci* in the Pacific Ocean. Nature, 228:341-342.

VINE, P.J. 1974. Effects of algal grazing and aggressive behaviour of the fishes *Pomacentrus lividus* and *Acanthurus sohal* on coral-reef ecology. Marine Biology (Berlin), 24: 131-136.

WALLACE, C.C. 1978. The coral genus *Acropora* (Scleractinia: Astrocoeniina: Acroporidae) in the central and southern Great Barrier Reef province. Memoirs of the Queensland Museum, 18(2):273-319.

WANDERS, J.B.W. 1977. The role of benthic algae in the shallow reef of Curacao (Netherland Antilles). III. The significance of grazing. Aquatic Botany, 3:357-390.

WEBER, J.N., and P.M.J. WOODHEAD. 1970. Ecological studies of the coral predator *Acanthaster planci* in the South Pacific. Marine Biology (Berlin), 6:12-17.

WELLINGTON, G.M. 1980. Reversal of digestive interactions between Pacific reef corals: Mediation by sweeper tentacles. Oecologia, 47(3):340-343.

WELLS, J.W. 1956. Scleractinia; pp. 328-444. *In*: Moore, R.C., ed., Treatise on Invertebrate Paleontology, Coelenterata. Lawrence: University of Kansas Press.

WELLS, J.W. 1972. Notes on Indo-Pacific scleractinian corals. Part 8. Scleractinian corals from Easter Island. Pacific Science, 26(2):183-190.

WENNER, A.M. 1971. The bee language controversy: An experiment in science. Boulder, Colorado: Educational Programs Improvement Corporation, 109 pp.

WEYGOLDT, P. 1977. Communication in crustaceans and arachnids. pp. 303-333. *In*: Sebeok, T.A., ed., How Animals Communicate. Bloomington: Indiana University Press.

WRIGHT, H.O. 1968. Visual displays in brachyuran crabs: Field and laboratory studies. American Zoologist, 8: 655-665.

NUTCRACKERS AND PINES: COEVOLUTION OR COADAPTATION?

Diana F. Tomback

Department of Biology
University of Colorado
Denver, Colorado

Since mutualistic associations are less common in the temperate zone than in tropical habitats, of particular interest are the interactions of seven North American and Eurasian species of Pinus *(subsections* Cembroides, Strobi, *and* Cembrae*) and the two species of* Nucifraga *(the "nutcrackers" of the* Corvidae. *From the interaction a nutcracker gains an energy-rich, efficiently-harvested food of storable form, and a pine gains a disperser with dissemination effected through the seed storing habits of the bird. Whereas the bird is the primary, if not the only, dispersal agent for the pine, nutcrackers will exploit other food sources.*

Morphological features adapted to the association include the long, sharp bill and sublingual pouch of the nutcrackers and the large, wingless seeds and seed retention in cones (in six species) of the pines. These nutcracker-pine interactions are considered coevolved, but information is lacking on whether the seed preferences of nutcrackers are genetically based. Nucifraga *probably originated in Eurasia, and it is likely that a form crossed the Bering Strait to North America before or during the Miocene. An evaluation of the evolutionary history of each nutcracker-pine relationship suggests that* N. caryocatactes *and the Eurasian* Cembrae *pines are most likely coevolved, and possibly also* N. columbiana *and* P. albicaulis. *Because the North American nutcracker extended its range to include* P. flexilis *and* P. edulis, *these associations are most likely coadapted.*

INTRODUCTION

A mutualistic interaction between two species is a reciprocally beneficial association. As May (1981) indicates, the interaction should increase the carrying capacity of each species (K_1 and K_2) relative to the population size of the associated species (N_1 or N_2): $K_1 \rightarrow K_1 + \alpha N_2$ and $K_2 \rightarrow K_2 + \beta N_1$ where α and β are interaction coefficients. Theoretically, however, mutualistic relationships are less stable than other kinds of species interactions and, following disturbance, take longer to regain equilibrium (May, 1973, 1981). According to Farnworth and Golley (1974), mutualistic interactions are more common in tropical than in temperate zone habitats because the conditions are more uniform year-round. "Stable and predictable" conditions permit a long association between species (May, 1981).

Recent studies have described a widespread temperate zone mutualistic association between the two avian species of the genus *Nucifraga*, the "nutcrackers" of the family Corvidae, and at least seven species of the genus *Pinus*, family Pinaceae. The pines and their nutcracker associates occur in the montane regions of western North America and Eurasia.

The general features of all nutcracker-pine interactions are as follows: 1) The nutcrackers are food specialists with a primary diet of fresh or stored seeds of the "nutcracker" pines. Relative to other sympatric pines, the profitability (defined as net caloric intake per caloric cost of handling) of these "nutcracker" pine seeds is highest. 2) Nutcrackers may be the primary dispersal agents, if not only the only dispersal agents, of these pines. Dispersal is effected by means of the seed-storing behavior of the birds. Nutcrackers bury clusters of pine seeds, hereby referred to as "caches", throughout their habitat in sites suitable for germination.

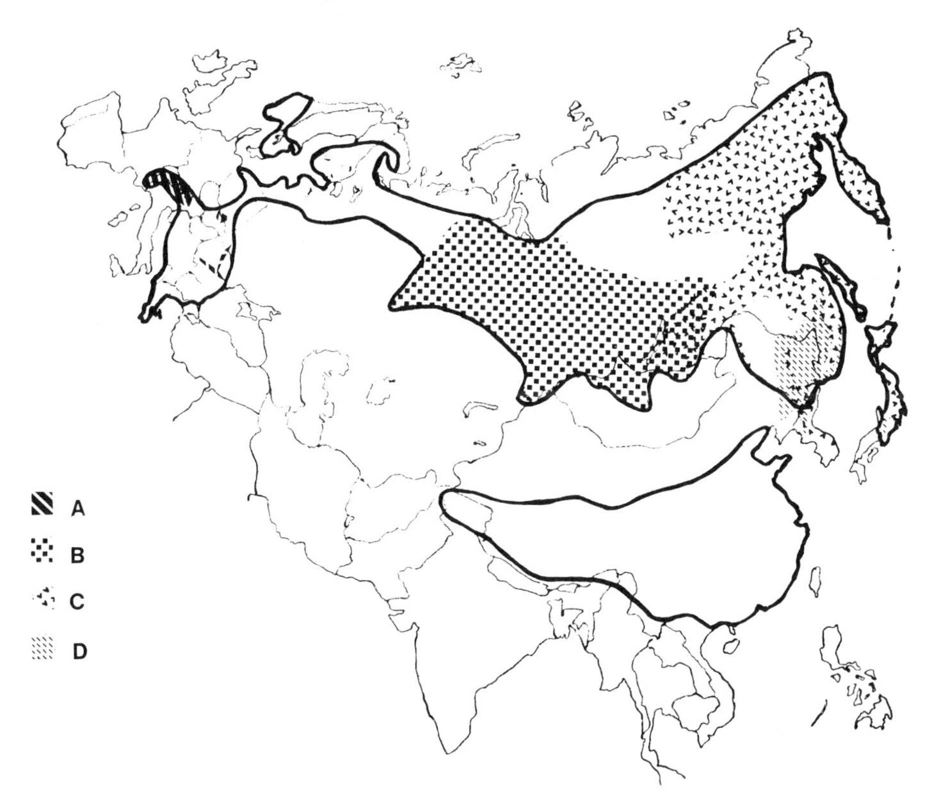

FIGURE 1. General distributions of (A) Pinus cembra, *(B)* P. sibirica, *(C)* P. pumila, *(D)* P. koraiensis, *and (heavy outline)* Nucifraga caryocatactes *throughout Europe and Asia. Data sources: Critchfield and Little (1966), Dementiev and Gladkov (1970), and Mattes (1978).*

Definition of Coevolution

As pointed out by Janzen (1980), the term "coevolution" has been too hastily and loosely applied to species' interaction, including mutualistic associations. Smith (1975) defined coevolution as "... the change in two or more species which are acting as selective forces on each other." Janzen's (1980) careful (but lengthy) definition requires that the specific changes effected in one species act as selective forces on the other species, and vice versa. Alternatively, the species in a mutualistic relationship can be coadapted rather than coevolved, i.e., the species can acquire adaptations in separate evolutionary histories which are fortuitously compatible.

Students of nutcracker-pine interactions suggest that the associations are coevolved (Balda, 1980a; Crocq, 1978; Lanner, 1980; Mattes, 1978; Tomback, 1978, 1982; Vander Wall and Balda, 1981). However, a rigorous definition of "coevolution" (e.g., Janzen, 1980) requires careful evaluation of each mutualistic system. I will describe the characteristics and taxonomic relationships of the species involved, provide ecological details from one particular association, and evaluate whether each nutcracker-pine association is coevolved or coadapted.

NUTCRACKERS

The genus *Nucifraga* consists of two species with similar behavior, morphology, and ecological requirements. The Eurasian Nutcracker (*N. caryocatactes*) is divided into ten subspecies (Dementiev and Gladkov, 1970) which range throughout montane regions of Europe and Asia (fig. 1). The Clark's Nutcracker (*N. columbiana*) (fig. 2) is monotypic and occurs in the higher montane regions of western North America, from southern Canada to northern Baja California and east as far as South Dakota (fig. 3). For both nutcrackers, the geographic distribution parallels (and in Europe, extends beyond) the occurrence of large-seeded pines (Critchfield and Little, 1966).

The long, sturdy bill of *Nucifraga* decurves slightly to

FIGURE 2. Clark's Nutcracker.

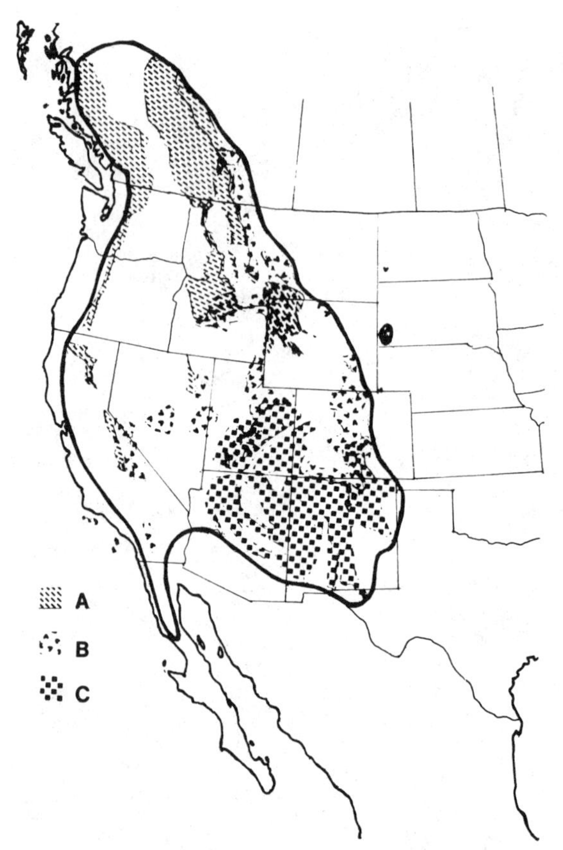

FIGURE 3. General distributions of (A) Pinus albicaulis, *(B)* P. flexilis, *(C)* P. edulis, *and (heavy outline)* Nucifraga columbiana *throughout North America. Data sources: Critchfield and Little (1966) and Peterson (1969).*

a point and is used to open pine cones and extract seeds. Also, unique to this corvid genus is the sublingual pouch, a sac-like modification of the floor of the mouth (described for the Eurasian species by Pallas, 1811, and Portenko, 1948, and for the New World species by Bock et al., 1973). Nutcrackers hold or transport pine seeds in the pouch, usually from harvested trees to storage sites. Another feature which facilitates the interaction with *Pinus* is the remarkable ability of nutcrackers to recall exact spatial locations of their seed caches, as demonstrated by recent field and laboratory experiments (Balda, 1980b; Tomback, 1980; Vander Wall, 1982).

A morphological consequence of a year-round pine seed diet may be the incubation patch of male nutcrackers (Mewaldt, 1952; Swanberg, 1956), a rare occurrence among passerine species (Bailey, 1952; Parkes, 1953). The patch may be an adaptation for early breeding; in both species, breeding begins in February or March when montane weather conditions are extreme and eggs must be constantly warmed (Mewaldt, 1956; Swanberg, 1956; Reimers, 1959). Early breeding may insure that young are adequately robust and experienced to become independent from parents, to store their own seeds the following fall, and to survive the montane winter (Vander Wall and Balda, 1977, 1981; Tomback, 1978). Crocq (1978) suggests that the incubation patch and behavior of the male allows the eggs to be well-attended while the female retrieves and eats seeds from her own stores.

PINES

Seven species of the genus *Pinus* have mutualistic interactions with nutcrackers (table 1). New World pines pinyon (*P. edulis*), limber (*P. flexilis*) and whitebark (*P. albicaulis*), are distributed through the western United States and southwestern Canada (fig. 3). Old World pines Swiss stone pine (*P. cembra*), Siberian stone pine (*P. sibirica*), Japanese stone pine (*P. pumila*), and Korean pine (*P. koraiensis*) are distributed through Eurasia (fig. 1). According to the classification of Critchfield and Little (1966), these pines belong to the subgenus *Strobus* or the "white pines" [subgenus *Haploxylon* of Shaw (1914) and Pilger (1926)]. The pinyon pines, including

Table 1. Nutcracker - Pine Mutualistic Interactions

Nutcracker Nucifraga *spp.*	*Pine*-Pinus *spp.*		*Geographic Range of pine*	*References*
N. columbiana	P. edulis	*pinyon*	*Southwestern United States*	*Vander Wall & Balda (1977)*
N. columbiana	P. flexilis	*limber pine*	*Western United States*	*Lanner & Vander Wall (1980)* *Tomback & Kramer (1980)* *Lanner (1980)*
N. columbiana	P. albicaulis	*whitebark pine*	*Western United States and Canada*	*Tomback (1978, 1981, 1982)* *Lanner (1980, 1982)*
N. c. caryocatactes	P. cembra	*Swiss stone pine*	*Switzerland and Austria*	*Holtmeier (1966)* *Crocq (1978)* *Mattes (1978)*
N. c. macrorhynchos	P. sibirica	*Siberian stone pine*	*Siberia*	*Bibikov (1948)* *Kondratov (1953)* *Reimers (1953)*
N. c. macrorhynchos *and* kamtschatkensis	P. pumila	*Japanese stone pine*	*N.E. Siberia and Japan*	*Kishchinskii (1968)* *Mezhenny (1961, 1964)*
N. c. macrorhynchos *and* japonicus	P. koraiensis	*Korean pine*	*S.E. Siberia, China, Korea and Japan*	*Mattes (1978)*

P. edulis, have traditionally been included in a distinct subgroup (Shaw, 1914) or with two Eurasian species (Pilger, 1926). Critchfield and Little (1966) grouped the pinyons alone in Subsection *Cembroides* of Section *Parrya*.

The taxonomic affinities of the "nutcracker" pines are important clues to the coevolution question. Unfortunately, much white pine taxonomy is yet unresolved. However, all proposed systems indicate that features facilitating seed dispersal by nutcrackers arose independently in two or three pine taxa. Much debate centers on the relationship of whitebark and limber pines. Shaw (1924) placed *P. flexilis* in group *Strobi* and *P. albicaulis*, *P. cembra*, *P. sibirica*, *P. pumila*, and *P. koraiensis* in group *Cembrae* [to which newly-described species were added by Mirov (1967)]. Pilger (1926) combined *P. flexilis* with the *Cembrae* pines, but Shaw (1924) and Critchfield and Little (1966) considered the *Cembrae* pines to be a distinct group with *P. flexilis* in subsection *Strobi*. Critchfield and Kinloch (pers. comm.) recently compiled data on experimental hybridization among species of the subgenus *Strobus*. Based on these data and other morphological features, Critchfield and Kinloch suggest that *P. albicaulis* and *P. pumila* be placed in subsection *Strobi* with *P. flexilis* and that subsection *Cembrae* consist of *P. cembra*, *P. sibirica*, *P. koraiensis*, and *P. armandii* (which occurs in China).

Traits in nutcracker-dispersed pines which increase the likelihood of seed harvest by nutcrackers are as follows:

1) In the majority of pine species, cone scale integument (spermoderm) forms a thin, woody membrane on each seed (Mirov, 1967). This seed "wing" enables wind dispersal of seeds when cones open. However, the seeds of nutcracker-dispersed pines are either wingless or nearly so, i.e., bearing functionless, narrow wing remnants (table 2). The "wingless" seeds are

Table 2. Characteristics of Nutcracker-Dispersed Pines in North America

	Pinus edulis *pinyon*	Pinus flexilis *limber*	Pinus albicaulis *whitebark*
Seeds	*large wingless*	*large wingless*	*large wingless*
Seed quality color signal	*yes*	*no*	*no*
Cones	*dehiscent seeds retained by flanges*	*dehiscent*	*indehiscent*
Cone orientation	*highly visible ca. horizontal*	*highly visible ca. horizontal*	*highly visible ca. horizontal*
Branch orientation	*upward*	*upward*	*upward*
Tree growth form	*single trunk*	*multiple trunks*	*multiple trunks*

Table 3. Energetic Value of Pine Seeds Without Coats.

Species	*cal per gram*	*Reference*
Nutcracker-dispersed		
P. edulis	7410	Little (1938) in Botkin and Shires (1948)
P. albicaulis	7320 {7716, 6925}	Tomback (1982) Lanner (1982)
P. cembra	7261 {7742, 6780}	Grodzinski and Sawicka-Kapusta (1970) Turcek (1967)
P. koraiensis	6953	Turcek (1967)
	$\bar{x}$ = 7236	
Wind-dispersed		
P. sylvestris	6482	Grodzinski and Sawicka-Kapusta (1970)
P. contorta	6786 (mean)	Smith (1968)
P. monticola	7408	Smith (1968)
P. nigra	4452	Turcek (1967)
P. ponderosa	7558	Smith (1968)
	$\bar{x}$ = 6537	

unlikely to be disseminated by wind, the trees may conserve some energy by producing little or no wing material, and nutcracker foraging efficiency is increased. Nutcrackers must remove the wings from wind-dispersed pine seeds before placing seeds in the sublingual pouch, an effort that slows their harvesting rate (Tomback, 1978).

2) The energetic content of seed endosperm is consistently high for nutcracker-dispersed pines in relation to sympatric wind-dispersed pines (table 3), although the difference is not significant (Mann-Whitney U test, P = .28, one tailed). Possibly, seeds preferred by nutcrackers have a high fat content, which may account for the high energy value, but few data are available to test this. According to Botkin and Shires (1948), edible portions of *P. edulis* and *P. flexilis* seeds contain about 62% and 52% fat, respectively, whereas the edible portion of singleleaf pinyon *P. monophylla* seeds, which is not a nutcracker-dispersed pine in the eastern Sierra Nevada (Tomback, 1978), contains 23% fat.

3) The seeds of nutcracker-dispersed pines are generally larger and heavier than those of wind-dispersed pines (table 4). The size difference between the two groups is highly significant (Mann-Whitney U test, P = .00023), and, in fact, the smallest nutcracker-dispersed seeds--those of *P. pumila*--are much heavier than the overall mean for wind-dispersed seeds (10,800 vs. 33,560 seeds per lb.). The seeds of nutcracker-dispersed pines are also significantly heavier than the seeds of wind-dispersed white pines (Mann-Whitney U test, P = .009). Large seed size increases nutcracker foraging efficiency (higher profitability of food items) and may be fundamental to nutcracker food preferences. The large seeds of nutcracker-dispersed pines probably evolved initially as an adaptation increasing seed germination success and seedling

Table 4. Seed Weights (Pinus *spp.*). [*Data from Table 6 in Krugman and Jenkinson (1974)*].

Species	*Cleaned seeds per pound*	*Species*	*Cleaned seeds per pound*	*Species*	*Cleaned seeds per pound*
NUTCRACKER-DISPERSED+		*WIND-DISPERSED**			
P. edulis	1900	P. aristata+	18,100	P. pinaster	10,000
P. flexilis	4900	P. attenuata	25,400	P. ponderosa	12,000
P. albicaulis	2600	P. balfouriana+	16,900	P. pungens	34,200
P. cembra	2000	P. banksiana	131,000	P. radiata	13,300
P. sibirica	1800	P. brutia	9,100	P. resinosa	52,000
P. pumila	10,800	P. canariensis	4,200	P. rigida	61,700
P. koraiensis	820	P. caribaea	31,000	P. roxburghii	5,600
		P. clausa	75,000	P. serotina	54,000
$\bar{x} = 3,546 \pm 3,438$ (S.D.)		P. contorta	135,000	P. strobus+	26,500
		P. coulteri	1,400	P. sylvestris	75,000
		P. densiflora	52,000	P. taeda	18,200
		P. echinata	46,300	P. thunbergiana	34,000
		P. elliotii	13,500	P. virginiana	55,400
		P. engelmannii	10,000	P. wallichiana	9,100
		P. glabra	46,000		
		P. halepensis	28,000		
		P. heldreichii	21,000		
		P. insularis	27,000		
		P. jeffreyi	3,700	* $\bar{x} = 34,388 \pm 30,493$ (S.D.)	
		P. lambertiana+	2,100		
		P. leiophylla	40,000	*+ $\bar{x} = 13,260 \pm 10,583$ (S.D.)	
		P. merkusii	18,200		
		P. monticola+	27,000		
		P. mugo	69,000		
		P. muricata	46,800		
		P. nigra	26,000		
		P. palustris	4,900		
		P. patula	52,600		

* *Other wingless or nearly wingless-seeded pines are not included. Some of these other pines may be dispersed by birds and/or rodents.*

\+ *Pines of subgenus* Strobus.

survival in harsh environments (i.e., arid and montane) (Baker, 1972; Smith, 1975), and nutcracker seed selectivity (Tomback, 1978; Tomback and Kramer, 1980) added additional pressure.

Seed size is controlled by the relative energy allocation between seeds and protective cone tissues by the parent. Smith (1975) discusses how strict seed predators, as opposed to seed dispersers which are also predators, select for a higher proportion of reproductive energy in protective tissues rather than in seeds. For example, seeds are only about 1% of the weight of the lodgepole pine (*P. contorta*) cones where red squirrels (*Tamiarsciurus hudsonicus*) are major predators (Smith, 1975). Using Smith (1970) data, Lanner (1982) calculated that the energetic value of the edible contents of the seeds of ponderosa pine (*P. ponderosa*) were about 8% and of lodgepole pine 1.6% the total value of cones plus seeds. In contrast, Lanner (1982) determined that the seeds of whitebark pine represent 17% to 46% ($\bar{x}$ = 29%) of the total energetic value of cones (plus seeds), and the edible portion of seeds (kernels) represents 5% to 29% ($\bar{x}$ = 17%) of the total.

4) For six of the seven nutcracker-dispersed pines, ripe seeds are retained in the cones (see table 2 for the exception); this prevents seeds from being dislodged by wind, increases the likelihood of nutcracker dispersal of seeds, and increases nutcracker foraging efficiency. In contrast, for wind-dispersed pines (except those species with fire-adapted, serotinous cones), cone scales open and release ripe seeds.

The cones of pinyon pine are dehiscent, i.e., scales open when cones are ripe (table 2). However, the ripe seeds are embedded in hollows on the cone scales, held in place by flanges, and are not dislodged after opening (Vander Wall and

Balda, 1977).

The cones of whitebark pine (table 2) and of the other *Cembrae* pines are classified as indehiscent (Shaw, 1914), i.e., scales do not open when cones ripen, although the scales may separate slightly (Shaw, 1914; Tomback, 1981; Lanner, 1982). The seeds of *Cembrae* pines are held in cones until nutcrackers and other seed predators tear the scales off. Lanner (1982), attributes the indehiscence of *P. albicaulis* cones to the absence of tracheid strands on the adaxial scale surface. In dehiscent species, these strands shrink and pull scales away from the cone axis, exposing the seeds (Harlow et al., 1964).

The cones of limber pine are dehiscent (table 2) and, consequently, seeds may be dislodged by wind soon after cones open. Nutcrackers foraging among open limber pine cones must search carefully through scales for seeds (Tomback and Kramer, 1980), and thus forage less efficiently.

5) The coat color of pinyon seeds is dark brown on edible seeds and light brown on aborted seeds, and nutcrackers appear to use color to discriminate between good and bad seeds (Vander Wall and Balda, 1977).

6) Whereas in most pines branches are horizontal to tree trunks, in pinyon, limber, and whitebark pine, they are upward-oriented, almost vertical, giving the trees a shrubby appearance (Vander Wall and Balda, 1977; Tomback, 1978; Lanner, 1982) (table 2). However, the branches of the Old World *Cembrae* pines appear to be horizontal although *P. pumila* occurs only as a low, mat-like form. While in wind-dispersed pines cones are usually downward-pointed on horizontal branches, all the cones of pinyon and whitebark pine (fig. 4A) and over half the cones of limber pine (Vander Wall and Balda, 1977; Tomback, 1978; Lanner, 1980 and 1982) are horizontally-oriented near the tips of vertical branches (fig. 4B). This cone and branch orientation appears to render the cones highly visible

A
B

to a flying bird and provides easy access to seeds (Vander Wall and Balda, 1977; Tomback, 1978; Lanner, 1980, 1982). Tomback (1978) describes how a nutcracker perches on a whitebark pine cone while harvesting seeds or on an adjacent cone of the same branch, thus obtaining a secure perch.

7) Unlike the cones of most wind-dispersed pines, the cones of all nutcracker-dispersed pines are not armed with sharp spines at the umbo of each scale. Spines presumably discourage vertebrate seed predators from harvesting seeds from ripe and unripe cones.

8) Finally, there is an interesting feature of nutcracker-dispersed pines which does not expedite seed harvest by nutcrackers, but which does appear to be a consequence of the "nutcracker" mode of seed dispersal. A common growth form of whitebark and limber pine is a multi-trunked tree (fig. 5)(Clausen, 1965; Weaver and Dale, 1974; Woodmansee, 1977; Lanner and Vander Wall, 1980; Lanner, 1980; Tomback, 1982). Sudworth (1908) described the growth form of whitebark pine as "often in clusters of from 3 to 7 trees, as if growing from the same root." Clausen (1965) surveyed 6000 whitebark pine trees in the eastern Sierra Nevada and found that over 70% were of the multi-trunked form with 5 and 6 per "tree" most frequent. Of the 1270 whitebark pine "stems" surveyed by Lanner (1980) in western Wyoming, 47% occurred with others in clusters of 2 to 8. The frequency of the multi-trunked form is much lower in other conifers (Clausen, 1965; Lanner, 1980).

Lueck (1980) excavated several young multi-trunk whitebark pines in the Cascade Range and found that all trunks

---

FIGURE 4. A. Cones of P. albicaulis. *B. Cones of* P. flexilis.

were separate. Thus, the seedlings of both whitebark and limber pine show a remarkable tolerance of crowding. Each "trunk" of a multi-trunked tree may arise from a single seed of a nutcracker cache (Tomback, 1982; Lanner, 1980). Therefore, a consequence of nutcracker seed dispersal is a clumped population dispersion pattern for these pine species.

ECOLOGICAL RELATIONSHIP BETWEEN CLARK'S NUTCRACKER AND WHITEBARK PINE

Details of the association between Clark's Nutcracker and whitebark pine should clarify the nature of mutualistic nutcracker-pine interactions. From 1973 to 1979 I investigated various aspects of the behavior and ecology of Clark's Nutcracker in relation to *P. albicaulis* on the south-central east slope of the Sierra Nevada range in Mono and Madera Counties, California. Since whitebark pine is restricted to subalpine elevations (Sudworth, 1908), primary study areas were between ca. 2700 and 3100 m and included Minaret Summit and Red's Lake on the west slope of Mammoth Mountain and the meadows and adjacent slopes at Tioga Pass (Tomback, 1978 and 1982).

Along the east slope, I have observed nutcrackers harvest and store seeds of *P. albicaulis*, *P. flexilis*, single leaf pinyon, and Jeffrey pine (*P. jeffreyi*). North of Lake Tahoe where Jeffrey pine and its close relative ponderosa pine are sympatric, nutcrackers may also harvest and store seeds of the latter pine. Nutcrackers probably use the seeds of these five species because the seeds are large relative to

---

FIGURE 5. Multi-trunked growth form of P. albicaulis.

other sympatric conifers (and wingless in the case of three species), the trees are locally abundant, and the sequence of cone ripening allows sequential utilization of two or more species (Tomback and Kramer, 1980). However, nutcrackers harvest and store whitebark pine seeds in preference to those of the other pine species, probably because whitebark pine cones ripen earlier (Tomback, 1978; Tomback and Kramer, 1980).

Annual Cycle of Clark's Nutcracker

The following information is summarized from Tomback (1978). The described events varied in timing by several weeks each year, depending in part on winter snowpack and other weather conditions. In summer at subalpine elevations, nutcrackers began to harvest seeds from partially ripe white-bark pine cones any time from the second half of July to the beginning of August. During this period, juvenile nutcrackers were still dependent on their parents, who periodically fed them pieces of unripe seeds. Adults and newly-independent juvenile nutcrackers began to store whitebark pine seeds from about the end of August to the second week of September. The onset of storage activities each summer apparently depended on sufficient ripeness of cones. The rate at which nutcrackers could extract seeds from ripe cones was significantly faster than that from unripe or partially ripe cones. Ripe seeds had hard coats which preserved them in storage, whereas the unripe seed coats were soft.

A nutcracker storing seeds hopped along the ground until it located a good site where it would dig a shallow trench with sideswiping motions of its bill. One by one, seeds were removed from the sublingual pouch and placed in a small cluster in the trench. The trench was then covered with soil, again by means of bill sideswipes. In areas with gravel or pumice substrate, the seeds were merely pushed one at a time in place with the bill tip. After making a cache, the nutcracker moved on, usually burying several other caches in the same area. Nutcrackers selected seed storage sites throughout the forest terrain, particularly near whitebark pine trees. They especially favored steep, south-facing slopes. Such sites accumulated minimal snow pack, and consequently, facilitated access to caches. In addition to storing whitebark pine seeds at subalpine elevations, nutcrackers sometimes transported seeds down to the Jeffrey pine belt (ca. 2100 m) and stored them in south-facing slopes.

Some nutcrackers continued to harvest and store *P. albicaulis* seeds until mid-October, by which time most cones were depleted and search time increased greatly. By mid-September, many nutcrackers migrated down to the Jeffrey pine belt where they began to harvest and store seeds, primarily of Jeffrey pine and also of singleleaf pinyon. Harvest and storage of Jeffrey pine seeds began as early as mid-September and as late as mid-October. Similarly, in the southern Sierra Nevada, by mid-September many nutcrackers ceased storing whitebark pine and began storing *P. flexilis* at lower elevations (Tomback and Kramer, 1980). In most years, storage of Jeffrey pine seeds probably ended in November.

During winter months, the majority of nutcrackers remained in the Jeffrey pine belt, although some were transient in the higher elevations. In the Jeffrey pine belt, nutcrackers foraged through cones on trees for remaining seeds and retrieved seeds from lower stores. Nesting began in March and April, or later at high elevations. Throughout spring, nutcrackers became more abundant at subalpine elevations where snow melt allowed access to seed stores.

By late June only a small number of nutcrackers remained

in the Jeffrey pine belt, whereas most first-year birds (non-breeding) and family groups had migrated up to subalpine elevations. At these higher elevations, until the birds began to harvest the new crop of whitebark pine cones in mid-summer, the principal foods were whitebark pine seeds retrieved from stores and insects. Throughout summer months at subalpine elevations, particularly after rainy periods, clusters of young seedlings appeared in sites similar to those used for seed storage. Thus, it appeared that nut-crackers were placing their caches in sites favorable to whitebark pine seed germination.

The Nutcracker as a Seed Disperser

The benefits gained by Clark's Nutcracker as a predator of whitebark pine seeds include an efficiently-harvested, energy-rich food of a form easily stored. It appears that whitebark pine benefits, in turn, from high-quality seed dispersal effected by Clark's Nutcracker. Details presented in Tomback (1982) are summarized below.

1) *Viability of cached seeds.* Experimental and empirical evidence suggests that the seeds buried by nut-crackers in caches are germinant. I compared the viability of twenty whitebark pine seeds removed from nutcracker caches within a few weeks of storage with the viability of twenty ripe seeds extracted from cones. Sliced seeds were immersed in a 0.1% solution of tetrazoluim chloride (see Machlis and Torrey, 1956). Nineteen seed embryos of each group stained red, indicating at least comparable viability. In fact, the seeds in nutcracker caches should be of greater overall viability than the seeds in a cone. Nutcrackers discriminate between good and inedible seeds, probably assessing seed quality by weight (Balda, 1980a), by "rattling" seeds between

their mandibles, i.e., a seed is moved up and down in the bill while the mandibles are rapidly opened and closed around it; and inedible seeds (e.g., aborted or insect-infested) are dropped (Vander Wall and Balda, 1977; Tomback, 1978).

2) *Number of seeds per cache: cluster size and growth form.* Based on observations of 17 nutcrackers burying 54 whitebark pine seed caches, I determined that nutcrackers bury from 1 to 15 seeds per cache with a mean, standard deviation, and median of 3.7, $\pm$ 2.9, and 3 seeds, respectively. Relatively small seed caches reduce competition for space, moisture, and light during germination and subsequent seedling growth. A consequence of small caches and tolerance of crowding may be reduced reproductive loss to whitebark pine from nutcracker mediated seed dispersal, resulting in the clumped tree distribution of the multi-trunked growth form. The frequency distribution of seeds per cache generally corresponds with that for the number of young seedlings per cluster appearing in the Mammoth Mountain study area (fig. 6), allowing for some non-germinating seeds or early seedling mortality. This is predicted from a cause and effect relationship between nutcracker caches and the multi-trunked growth form.

3) *Cache depth and cache site quality.* Seed depth is an important aspect of dispersal quality, and optimal seed depth is proportional to seed size (Hoffman, 1924). When nutcrackers retrieve whitebark pine seed caches from mineral soil, their excavations remain. Measurements of 26 such excavations at Tioga Pass indicated that seeds were buried not more than 1.0 to 3.0 cm deep [$\bar{x}$ = 2.0 $\pm$ 0.77 (5.0) cm]. In nursery practice, large seeds, such as *P. albicaulis*, are generally covered with about 1.25 cm of soil (Krugman and Jenkinson, 1974) but can germinate at lower depths, depending on soil and moisture conditions (Hoffman, 1924; Arndt, 1965

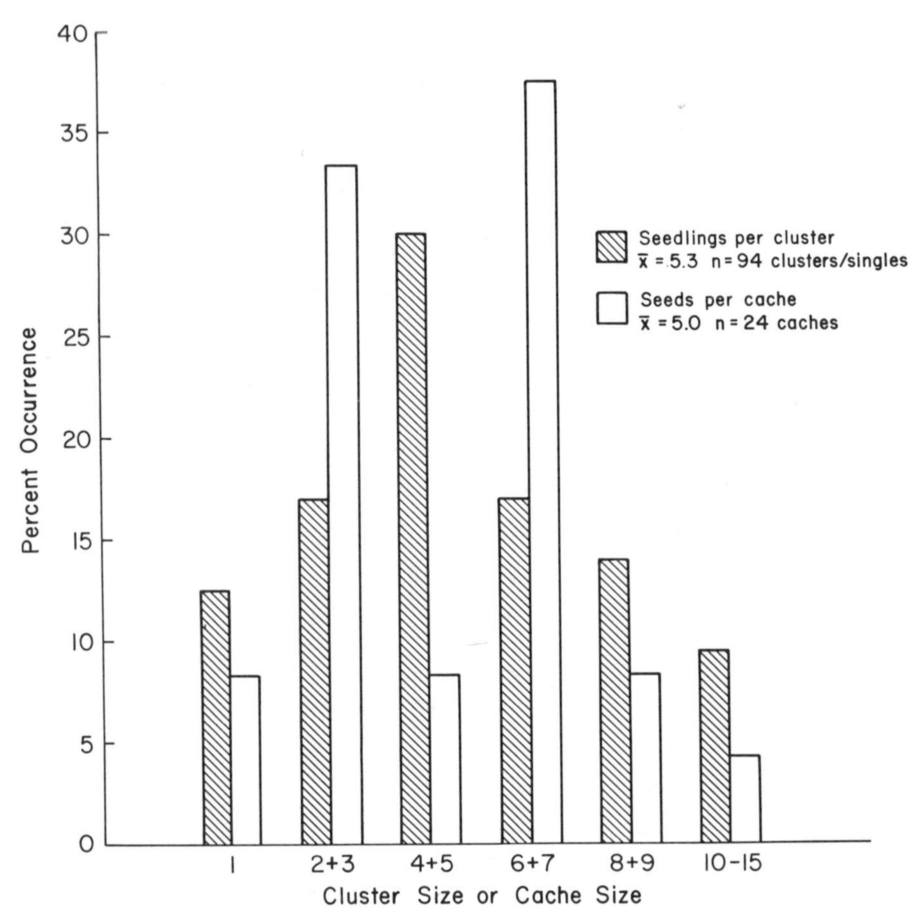

FIGURE 6. Comparison of frequency distributions of number of seeds per cache and number of seedlings per cluster.

in Koller, 1972). Consequently, the depths at which nutcrackers bury *P. albicaulis* seeds seem favorable for germination.

Another critical factor is the quality of cache sites selected relative to seed germination requirements. Table 5 is based on observations of 52 nutcrackers making 80 whitebark pine seed caches on Mammoth Mountain. Nutcrackers buried their seed caches in well-drained substrates, including soil,

Table 5. The Relative Frequency of Micro-Habitats Selected for Seed Storing by Nutcrackers (From Tomback, 1978).

General Type	Micro-habitat	Frequency	Subtotal
A. next to objects (within 1.5 m)	trees	.390	.615
	rocks	.120	
	fallen trees or branches	.080	
	plants	.025	
B. open substrate	pumice	.175	.265
	forest litter	.090	
C. in trees	roots	.090	.120
	holes, cracks underbark	.030	

forest litter, gravel, and pumice (the latter predominated on volcanic Mammoth Mountain). I never observed nutcrackers use moist substrate, a behavior which may prevent seed spoilage.

Pumice and gravel are poor seedbeds since they do not hold water. Also poor is forest litter, since the cellulose walls will trap and withhold moisture or heat up to potentially damaging temperatures for young seedlings (Baker, 1950).

Nutcrackers placed a relatively high frequency of caches (ca. 61.5%) next to objects such as rocks, fallen trees, fallen branches, a variety of perennial plants, and particularly tree trunks. In the latter case (ca. 39% of cache sites), the substrate was usually forest litter, a poor seedbed. Since *P. albicaulis* is classified as shade intolerant (Baker, 1950), seedlings originating next to trees may later lack light, moisture, or space. Nutcrackers also buried caches (ca. 26.5%) in open spaces more than 1.5 m from objects. Miscellaneous locations in trees provided a small proportion of cache sites for nutcrackers (ca. 12.0%). Seeds were tucked among roots, in cracks and holes, and under bark. These tree sites cannot support germination or seedling growth, and seeds stored there represent reproductive loss to *P. albicaulis*, as do the seeds transported and stored by nutcrackers below subalpine elevations.

Nutcrackers also stored seeds among rock rubble and in the fissures and cracks on exposed rock faces. Occasionally, whitebark pines did grow from such precarious locations. At Tioga Pass, nutcrackers buried seeds on the dry rocky rises

---

FIGURE 7. Newly-germinated seedlings originating from a nutcracker cache next to Eriogonum umbellatum.

which bordered the meadowland and small lakes and supported stands of *P. albicaulis*.

Some indication of seed germination potential of cache sites of general type A and B (table 5) was obtained by comparison with the microhabitat characteristics of 94 young seedling clusters in the Red's Lake study area on Mammoth Mountain. Apparently, seeds germinated at higher frequency than expected in open pumice, particularly after several days of rain. Seedling clusters next to sulphur flower plants (*Eriogonum umbellatum*) also occurred at a higher frequency than suggested in table 5 (plants), perhaps because the low, dense plant base provided coolness and moisture during the heat-sensitive stage of seedling growth (fig. 7). Seedling clusters near trees were observed at half the frequency such sites were used. In total, about 40% of the sites at subalpine elevations selected by nutcrackers for seed caching were potentially favorable for seedling growth; and whereas other sites were poor, some could still support seedlings.

4) *Seedling survival*. For whitebark pine population recruitment to occur, the sites selected by nutcrackers for seed caching must favor not only germination but also seedling survival. Mortality rates are generally high for the first few years of the tree's life under natural conditions (Baker, 1950). In 1975 and 1976 I conducted a short-term study of survival rates on two sites with seedling clusters, presumably originated from nutcracker caches (*Site 1* - 54 seedlings in 12 clusters, *Site 2* - 31 seedlings in 14 clusters). The first-year survival rate did not differ significantly between the two sites (65% and 41%, respectively) and the combined rate was 56%. Site 1 was examined again in 1979 after five growing seasons. Seedlings survived in 6 of the original 12 clusters (50%) with an overall survival rate of 26% of the

original number.

Also, in 1975, I encountered more than 600 seedlings likely to have originated in nutcracker caches in the Red's Lake area. More than 30% of these seedlings were still present by September 1976. All these seedling survival rates are similar to those obtained in reforestation efforts and nursery practice (Baker, 1950).

5) *Quantity of seeds stored.* Another critical factor pertaining to the quality of *P. albicaulis* dispersal effected by nutcrackers is the quantity of seeds stored in relation to the quantity of seeds actually retrieved. Estimates and empirical data on: 1) daily harvest duration, 2) number of seeds per pouchload, 3) time to fill a pouch, store seeds and return, 4) the number of days whitebark pine seeds were stored each year, and 5) the number of storage trips to lower elevations per bird, suggest that in a good cone crop year, each nutcracker stores about 32,000 whitebark pine seeds or 8500 caches at subalpine elevations. The energetic requirements of nutcrackers during the period of whitebark pine seed cache retrieval, i.e., April through July, were calculated with data from laboratory determinations of nutcracker metabolic rates (Laudenslager and Tomback, in preparation). From April through July, a nutcracker requires from 6000 to 10,000 seeds or 1600 to 3000 caches, which suggests that one bird stores between 3 and 5 times as many seeds as it needs.

However, since stored whitebark pine seeds are fed by breeding birds to juveniles, the excess each year actually depends on the cone crop size the previous fall in relation to nutcracker population size and demographics. For example, I estimated the surplus seeds stored one year by the nutcracker population in the Red's Lake area to be 45%, or about 350,000 superfluous seeds in 95,000 caches. Of these, some would be lost to rodents and spoilage, and for others, reproductive potential would be decreased by storage in unfavorable sites.

In fact, seed germination may begin in late spring or early summer regardless of how many caches have not been retrieved by nutcrackers. Consequently, the fact that nutcrackers use stored *P. albicaulis* seeds relatively late in the year increases the likelihood of germination of stored seeds.

6) *"Pioneering" effect.* In my study areas nutcrackers transport *P. albicaulis* over distances up to 12 km to storage sites. Vander Wall and Balda (1977) observed Clark's Nutcrackers transport seeds up to 22 km. As previously discussed, nutcrackers will store seeds at some distance from cone-bearing pines and on high rock ledges and crevices. Although the germination potential of seeds stored under such harsh conditions is low, stunted and twisted whitebark pines are occasionally encountered in such isolated and precarious sites. Consequently, an additional benefit to the pine from seed dispersal by nutcrackers appears to be seed dissemination to locations where there are few or no cone-bearing pines. The cache site preferences of nutcrackers may be an important factor in the distribution of whitebark pine throughout the subalpine habitat and in defining the limits of its occurrence.

A good illustration of the "pioneering" effect of nutcracker seed dissemination is post-fire regeneration. In August and September 1979, I searched on the top west slope of Cathedral Peak, elevation 3300 m, in Yosemite National Park, for signs and possible sources of regeneration in a 2 hectare tract burned by a lightning-strike fire in August 1975. The pre-fire forest was exclusively krummholz whitebark pine, the dwarf growth form described by Clausen (1965), which occurs under harsh growing conditions (Tranquillini, 1979). Data on pre-fire tree growth form and the occurrence and size of seedling clusters were gathered from five, parallel transects across the burn.

Throughout most of the burned tract, the fire left charred but intact tree skeletons. This afforded a unique opportunity to study the growth form of krummholz *P. albicaulis*, as the branches and foliage on live trees are usually too thick to determine tree structure. Many of the trees appeared multi-trunked, i.e., clusters of individuals as described for the erect growth form of this species. Although the "trunks" were often fused at the base, there were lines of separation between trunks and distinct wood grain patterns. The number of trunks ranged from 1 to 5 per tree, with a mean, standard deviation, and median of 2.2, ± 1.08, and 2, suggesting that at least some trees originated from nutcracker caches.

A total of 31 solitary seedlings and seedling clusters, five years of age or younger (estimated by height), were encountered along the transects. The number of seedlings per cluster ranged from 1 to 4 with a mean, standard deviation and median of 1.7, ± 0.96 and 1. All seedlings in clusters were separate down to the root, again suggesting an origin from nutcracker caches.

In September, nutcrackers were harvesting and storing seeds from the erect *P. albicaulis* on the lower east and west slopes of Cathedral Peak. On two occasions a nutcracker with an inflated sublingual pouch flew up from the lower slope into the burned area and buried caches. Thus, Clark's Nutcrackers were the likely source of regeneration for the Cathedral Peak burn. Similar nutcracker-mediated post-fire regeneration was observed for limber pine by Lanner and Vander Wall (1980).

The role of the Eurasian Nutcracker in the pioneering tendencies of *P. cembra* and *P. sibirica* have been well-documented by European workers. Of particular interest is the maintenance by nutcrackers of timberline at the highest possible elevations for these pine species, despite yearly fluctuations in weather conditions (e.g., Holtmeier, 1966; Crocq, 1978; Mattes, 1978).

Alternative Dispersal Agents

It is important to clarify whether Clark's Nutcracker is is the primary and most likely disseminator of whitebark pine seeds. The possible alternative dispersal agents are other vertebrate species and eventual cone disintegration.

In my study areas, no other avian species stored whitebark pine seeds, and deer mice (*Peromyscus maniculatus*)--notorius seed predators--were unlikely to gnaw into and remove seeds from intact *P. albicaulis* cones. However, Douglas squirrels (*Tamiasciurus douglasi*) cut down ripe *P. albicaulis* cones for stockpiling in middens, and chipmunks (*Eutamias* spp.) harvested seeds from *P. albicaulis* cones. Little whitebark pine population recruitment occurs from Douglas squirrel middens, as the situation is unsuitable for germination and seedling growth (Finley, 1969). Chipmunks are known to store food in late summer and autumn (Tevis, 1953), but many food caches are probably underground in burrows (Gordon, 1943). West's (1968) suggestion that some chipmunk surface caches may germinate and produce trees needs confirmation. Even if chipmunks were responsible for some whitebark pine propagation, they are less effective because their caches are large (West, 1968), they harvest seeds slowly, they are restricted to a small range, and they are severely outcompeted by nutcrackers and squirrels in the procurement of whitebark pine seeds.

Some experimental evidence suggests that unharvested *P. albicaulis* cones may abcise and disintegrate (Krugman and Jenkinson, 1974). Since this process may deposit many seeds in one site, it is likely to be less efficient than nutcracker dispersal. This, however, is a moot point because data gathered in fair and good cone crop years suggest that no intact cones remain on trees after nutcrackers, squirrels, and chipmunks cease foraging in fall (Tomback, 1981; Lanner, 1982). Disintegration may well offer an "escape route" for seeds in overlooked cones and may be, along with indehiscence, an adaptation for a dispersal system primarily dependent on nutcrackers.

COEVOLUTION OR COADAPTATION?

The Evolution of Large, Wingless Seeds

Whether nutcrackers and pines are coadapted or coevolved, effective nutcracker-mediated seed dispersal exerts selection pressure for large, wingless seeds. Potential candidates for the transition from wind dispersal (equals ancestral condition, Lanner, 1980) to corvid dispersal are those pines which, either because of climatic changes or range extension, encounter a stressful environment, i.e., conditions of aridity or high altitude (Baker, 1972). The seed size of these species may increase to provide larger food reserves, which increases the survival rate of young seedlings but decreases the effectiveness of dispersal by wind (Baker, 1972). If seed-storing corvids, including nutcrackers, are sympatric with such pine species, the possible alternative subsequent events are as those diagrammed in fig. 8.

If the germination requirements of a pine are incompatible with corvid dispersal, a bottleneck is encountered

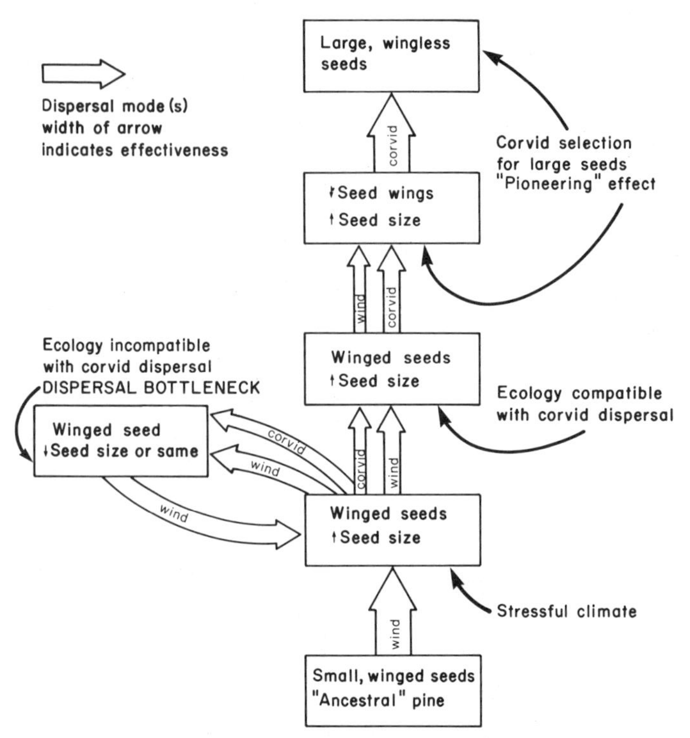

FIGURE 8. A model for the evolution of large, wingless seeds and corvid-mediated seed dissemination.

and seed size either decreases or remains the same, and wind dispersal is maintained. However, if the germination requirements are compatible with corvid storage sites and modes, then seed traits promoting corvid dispersal may evolve (fig. 8).

Specificity of Interactions

The bill structure and sublingual pouch of nutcrackers can be used for foraging on the seeds of any relatively large-seeded conifer and the seeds of other taxa as well. These features did not necessarily evolve only for large pine seeds. In fact, the geographical range of Clark's Nutcracker extends beyond those species it disperses. As previously mentioned, Clark's Nutcracker also takes seeds of the Jeffrey, single-leaf pinyon, and ponderosa pines and are known to eat the seeds of Douglas-fir (*Pseudotsuga menziesii*) and possibly of sugar pine (*P. lambertiana*) (Davis and Williams, 1957).

Six subspecies of the Eurasian Nutcracker occur outside the range of *Cembrae* pines (Mattes, 1978). Sympatric with these birds are several species of large-seeded pines, of which three have wingless seeds. In addition, various subspecies of the Eurasian Nutcracker are known to take spruce seeds (*Picea* spp.), walnuts (*Juglans* spp.) and hazelnuts (*Corylus* spp.) (Turcek and Kelso, 1968; Ali and Ripley, 1972).

The populations of the thick-billed Nutcracker (*N. c. caryocatactes*) in southern and central Sweden, where *P. cembra* does not occur, depend upon hazelnuts (*C. avellana*) as their primary food source. There, nutcrackers are territorial and gather hazelnuts from distant areas for storage. The stored nuts are the main food of nestlings, and clutch size is regulated by the hazelnut crop size of the previous autumn (Swanberg, 1981). No major differences in trophic structures between the populations of *N. c. caryocatactes* which depend on *P. cembra* and those which depend on *C. avellana* have been documented.

In contrast to the flexibility of nutcrackers in regard to food sources, of all the nutcracker-dispersed pines, only *P. edulis* has an effective alternative dispersal agent. Pinyon Jays (*Gynnorhinus cyanocephalus*), which are sympatric with nutcrackers in some montane areas, store large quantities of pinyon seeds and are important dispersal agents of the pine (Ligon, 1978). It is possible that seed drop and dissemination by deer mice, *Tamiasciurus* squirrels, and chipmunks, provide some alternative dispersal mechanisms for limber pine, whose cones are dehiscent and ripe seeds easily dislodge.

Although the "nutcracker" pines are essentially obligate mutualists, they exert some control over dispersal agents. Conifers are known to synchronize cone crop production over large geographic areas. That is, trees of the same species, and also of many species, will exhibit widespread cone crop failures or abundances (Bock and Lepthien, 1976). Ligon (1978) and Smith and Balda (1979) suggest that synchronized failures decrease populations of all seed predators and of the dispersers. Such major cone crop failures result in the mass exodus of both Clark's and Eurasian Nutcrackers from their montane habitats (Davis and Williams, 1957, 1964; Conrads and Balda, 1979, and references therein). Infrequent and unpredictable seed abundances swamp predators and dispersers, insuring that some pine reproduction can occur. It is probable that the availability, and utilization by nutcrackers, of alternative food sources has contributed stability to the nutcracker-pine association, enabling it to endure yearly fluctuations in seed crop size.

Coevolution or Coadaptation and Coercion?

Considering the diversity of seed sources that nutcrackers exploit, the possibility exists that nutcrackers prefer seeds of the "nutcracker" pines simply because these pines provide optimal foraging efficiency. In other words, attractive cone and seed traits "coerce" nutcrackers into concentrating their harvest and storage efforts on these species. Even if such traits gradually evolved in order to attract greater numbers of these effective dispersers, the relationship would be coadapted rather than coevolved because no change in the nutcracker need occur. However, if preferences for the cones and seeds of these species were genetically based in nutcrackers, the relationship would be considered coevolved.

It is not known whether specific seed preferences of nutcrackers are genetically based. Young nutcrackers begin to harvest and store seeds of these pines in their first summer, but seed preferences may be learned from other birds as nutcrackers tend to forage in small groups (e.g., Vander Wall and Balda, 1977; Tomback, 1978; Bibikov, 1948; Mezhenny, 1961). During an invasion of *N. c. macrorhynchos* into Bielefeld, Germany, following widespread cone crop failure of *P. sibirica*, Balda (1980, see also Conrads and Balda, 1979) observed yearling nutcrackers go directly to a feeder and take *P. sibirica* seeds. Balda thought it unlikely that the young birds had ever before seen *P. sibirica* seeds, suggesting an inherent recognition. This evidence is tentative at best since juveniles are sometimes fed intact seeds retrieved from caches of the previous year and might learn preference at an early age. Also, during cone crop failures a few trees are likely to produce cones, affording a learning opportunity for some individuals.

A second approach to the question of whether nutcracker-pine associations are coevolved or coadapted is a comparison of the evolutionary history of genus *Nucifraga* and each pine species. Turcek and Kelso (1968) mention as "prevalent theory" that an ancestral nutcracker form and *Cembrae* pine crossed the Bering Strait and eventually gave rise to *N. columbiana* and *P. albicaulis*. Lanner (1980) discusses this possibility in detail, pointing out that whereas there are four *Cembrae* pine species and 10 nutcracker subspecies in Eurasia, there is only one of each taxon in North America. Lanner (1980) suggests an Old World origin for both forms and states "...it seems highly probable that the *Cembrae* pines are the products of speciation from a common progenitor that was itself the product of nutcracker selection." If this scenario is correct, nutcrackers and *Cembrae* pines are probably coevolved.

The occurrence of a *Cembrae* pine and nutcracker in North America may well date back to the late Miocene. Brodkorb (1972) describes a fossil corvid humerus from a Miocene formation in Logan County, Utah, which resembles that of both Clark's Nutcracker and Pinyon Jay. However, fossils verified as both *N. columbiana* and *N. caryocatactes* date back only to the Pleistocene (Brodkorb, 1978). Of interest is the fact that the Cascade Range and Sierra Nevada began forming in the late Miocene while the Great Basin became more arid. Many species of pines described from the Miocene and Pliocene are similar to modern species (Mirov, 1967).

In their proposed revision of white pine classification, Critchfield and Kinloch (pers. comm.) suggest that whitebark pine is derived from limber pine and belongs in subsection *Strobi*. They suggest that *P. pumila* be transferred to *Strobi* from subsection *Cembrae* and that *P. armandii* be added to *Cembrae*. According to these new arrangements, cone indehiscence arose independently in two major taxa.

If these revised relationships are accurate, the whitebark pine-nutcracker association is likely to be coadapted, unless an inherent seed preference for *P. albicaulis* seeds can be demonstrated. In regard to the Eurasian species, nutcrackers may be coevolved with pines of both subsection *Cembrae* and *Strobi*.

The history and derivation of *P. flexilis* is better known. Limber pine is the northernmost pine of an intergrading series, beginning with *P. ayacahuite* in Central America and Mexico which grades into *P. strobiformis* in northern Mexico and finally *P. flexilis*. From *P. ayacahuite* to limber pine there is clinal variation in seed characteristics. The southernmost variety of *P. ayacahuite*, i.e., var. *ayacahuite*, has long narrow seed wings and small seeds, and the central variety *veitchii* has a larger seed and short, wide seed wings (Martinez, 1948). The northernmost variety *brachyptera* is considered synonymous with *P. strobiformis* by some workers (e.g., Critchfield and Little, 1966; Mirov, 1967) and has a large, essentially wingless seed. The seeds of *P. flexilis* are similar but somewhat smaller (Table 6 and fig. 2 in Krugman and Jenkinson, 1974).

Lanner (1980) points out that a multi-trunked form and vertical branches appear in some trees of var. *brachyptera* (= *P. strobiformis*), suggesting that clinal variation occurs in tree morphology as well. He further suggests that the clinal variation is the consequence of seed dispersal by corvids and that in arid habitats seed dissemination by several species of food-caching jays may improve reproductive success. The geography of the clinal variation may well indicate where climatic conditions are such that the evolutionary fitness of this pine "super" species is improved by corvid rather than wind dispersal.

If nutcrackers migrated into North America with a *Cembrae* pine, they probably encountered *P. flexilis* during range expansion into the southwest. It is possible that seed dispersal by nutcrackers enabled *P. flexilis* to invade more northern and subalpine regions. Without information on the southern boundary of nutcracker distribution in the Tertiary, it is safe to state that the association between Clark's Nutcracker and limber pine is coadapted rather than coevolved.

Of the pinyon pine species of subsection *Cembroides*, four occur in the United States and five have relictual distributions in the desert mountains of Mexico. According to Mirov (1967), the pinyon pines probably evolved in North America in the Oligocene from an ancestral white pine when the southwest became very arid. All nine pinyon pines are characterized by large, wingless seeds, which probably evolved in response to dispersal by one or more species of food-storing jays, in particular a Pinyon Jay form and also perhaps rodents. Nutcrackers probably encountered pinyon pine in the southwest and may be instrumental in the northern expansion of the range of this species. Alternatively, nutcrackers encountered *P. edulis* in its present distribution and included the seeds among its preferred food sources. In either case, the conservative approach is to consider the relationship with nutcrackers coadapted rather than coevolved.

In conclusion, the nutcracker-pine associations most certain to be coevolved rather than coadapted are between the Eurasian Nutcracker and several species of Eurasian pines, all currently in subsection *Cembrae*. If the affinity of *P. albicaulis* is also with *Cembrae* pines, then the association between Clark's Nutcracker and whitebark pine is coevolved as well.

ACKNOWLEDGEMENTS

R.M. Lanner, W.B. Critchfield and B.B. Kinloch kindly permitted me to read their unpublished works. Yan B. Linhart provided helpful and interesting comments on the manuscript. The Russian literature was translated by L. Kelso. Kathryn A. Kramer assisted me in the field in 1979. The post-fire regeneration study was funded by the U.S.D.A. Forest Service, Pacific Southwest Forest and Range Experiment Station, Berkeley, California, and my work in Yosemite National Park was coordinated by Jan W. van Wagtendonk, park Research Scientist. Support for manuscript preparation was provided by a University of Colorado Junior Faculty Development Award.

LITERATURE CITED

ALI, S., and S.D. RIPLEY. 1972. Birds of India and Pakistan, Vol. V. Bombay: Oxford University Press. 276 pp.

BAILEY, R.E. 1952. The incubation patch of passerine birds. Condor, 54:121-136.

BAKER, F.S. 1950. Principles of Silviculture. New York: McGraw-Hill Book Co. 414 pp.

BAKER, H.G. 1972. Seed weight in relation to environmental conditions in California. Ecology, 53:997-1010.

BALDA, R.P. 1980a. Are seed caching systems coevolved? *In* Nöhrig, R. (ed.), Acta XVII Congressus Internationalis Ornithologie.

BALDA, R.P. 1980b. Recovery of cached seeds by a captive *Nucifraga caryocatactes*. Zeitschrift für Tierpsychologie, 52:331-346.

BIBIKOV, D.I. 1948. On the ecology of the nutcracker. Trudy Pechorskogo-Ilychskogo Gosudarstvennogo Zapovednika, IV: 89-112. (In Russian).

BOCK, C.E., and L.W. LEPTHIEN. 1976. Synchronous eruptions of boreal seed-eating birds. American Naturalist, 110: 559-571.

BOCK, W.J., R.P. BALDA, and S.B. VANDER WALL. 1973. Morphology of the sublingual pouch and tongue musculature in Clark's Nutcracker. Auk, 90:491-519.

BOTKIN, C.W., and L.B. SHIRES. 1948. The composition and value of piñon nuts. New Mexico Experiment Station Bulletin, 344:3-14.

BRODKORB, P. 1972. Neogene fossil jays from the Great Plains. Condor, 74:347-349.

BRODKORB, P. 1978. Catalogue of fossil birds, Part 5 (Passeriformes). Bulletin Florida State Museum, Biological Sciences, 23:139-228.

CLAUSEN, J. 1965. Population studies of alpine and subalpine races of conifers and willows in the California high Sierra Nevada. Evolution, 19:56-68.

CONRADS, K., and R.P. BALDA. 1979. Überwinterungschancen Sibirischer Tannenhäher (*Nucifraga caryocatactes macrorhynchos*) im Invasionsgebiet. Bericht des Naturwissenschaftlichen Vereins Bielefeld, 24:115-137.

CRITCHFIELD, W.B., and E.L. LITTLE, JR. 1966. Geographic distribution of the pines of the world. U.S. Department of Agriculture, Miscellaneous Publication, 991: 97 pp. Washington, D.C.

CROCQ, C. 1978. Écologie du Casse-noix (*Nucifraga caryocatactes* L.) dans les Alpes françaises du sud: Ses relations avec l'Arolle (*Pinus cembra* L.). Unpublished dissertation, L'Université de Droit D'Économie et de Sciences D'Aix-Marseille. 189 pp.

DAVIS, J., and L. WILLIAMS. 1957. Irruptions of the Clark Nutcracker in California. Condor, 59:297-307.

DAVIS, J., and L. WILLIAMS. 1964. The 1961 irruption of the Clark's Nutcracker in California. Wilson Bulletin, 76: 10-18.

DEMENTIEV, G.P., and N.A. GLADKOV. 1970. Birds of the Soviet Union, Vol. V. Jerusalem: Israel Program for Scientific Translations. 957 pp.

FARNWORTH, E.G., and F.B. GOLLEY (eds.). 1974. Fragile ecosystems: Evaluation of research and applications in the neotropics. New York: Springer-Verlag.

FINLEY, R.B., JR. 1969. Cone caches and middens of *Tamiasciurus* in the Rocky Mountain region; pp. 233-273. *In* Jones, J.K. (ed.), Contributions in Mammalogy. Museum of Natural History, The University of Kansas.

GORDON, K. 1943. The natural history and behavior of the western chipmunk and the mantled ground squirrel. Oregon State Monographs, Studies in Zoology, 5:104 pp.

GRODZINSKI, W., and K. SAWICKA-KAPUSTA. 1970. Energy values of tree-seeds eaten by small mammals. Oikos, 21:52-58.

HARLOW, W.M., W.A. CÔTÉ, JR., and A.C. DAY. 1964. The opening mechanism of pine cone scales. Journal of Forestry, 62:538-540.

HOFFMAN, J.V. 1924. Natural regeneration of Douglas-fir in Pacific Northwest. United States Department of Agriculture Bulletin, 1200.

HOLTMEIER, F.K. 1966. Die ökologische Funktion des Tannenhähers in Zirben-Lärchenwald und an der Waldgrenze des Oberengadins. Journal für Ornithologie, 107:337-345.

JANZEN, D.H. 1980. When is it coevolution? Evolution, 34: 611-612.

KISHCHINSKII, A.A. 1968. Birds of the Kolyma Highlands. Pp. 100-109. (In Russian).

KOLLER, D. 1972. Environmental control of seed germination. *In* Kozlowski, T.T. (ed.), Seed Biology, Vol. II. New York: Academic Press.

KONDRATOV, A.V. 1953. On the restoration of the Siberian Cedar-pine in the wild by nest (cluster) sowing. Agrobiologiia, 3:161-164.

KRUGMAN, S.L., and J.L. JENKINSON. 1974. *Pinus* L. Pine; pp. 598-638. *In* Seeds of Woody Plants in the United States, U.S. Department of Agriculture, Forest Service. U. S. Department of Agriculture, Agriculture Handbook, 450.

LANNER, R.M. 1980. Avian seed dispersal as a factor in the ecology and evolution of limber and whitebark pines. *In*

Sixth North American Forest Biology Workshop, University of Alberta, Edmonton, Alberta.
LANNER, R.M. 1982. Adaptations of whitebark pine for seed dispersal by Clark's Nutcracker. Canadian Journal of Forest Research, 12:391-402.
LANNER, R.M., and S.B. VANDER WALL. 1980. Dispersal of limber pine seed by Clark's Nutcracker. Journal of Forestry, 78:637-639.
LIGON, J.D. 1978. Reproductive interdependence of Piñon Jays and piñon pines. Ecological Monographs, 48:111-126.
LUECK, D. 1980. Ecology of *Pinus albicaulis* on Bachelor Butte, Oregon. Unpublished M.A. thesis, Oregon State University.
MACHLIS, L., and J.G. TORREY. 1956. Plants in Action: A Laboratory Manual of Plant Physiology. W.H. Freeman and Co.
MARTINEZ, M. 1948. Los Pinos Mexicanos. Mexico City: Ediciones Botas.
MATTES, H. 1978. Der Tannenhäher (*Nucifraga caryocatactes* (L.)) im Engadin: Studien zu seiner Ökologie und Funktion im Arvenwald (*Pinus cembra* L.). Münstersche Geographische Arbeiten, Heft 2. Paderborn: Ferdinand Schöningh. 87 pp.
MAY, R.M. 1973. Qualitative stability in model ecosystems. Ecology, 54:638-641.
MAY, R.M. 1981. Models for two interacting populations; pp. 78-104. *In* May, R.M. (ed.), Theoretical Ecology, 2nd ed. Sunderland, Mass.: Sinauer Associates. 489 pp.
MEWALDT, L.R. 1952. The incubation patch of the Clark Nutcracker. Condor, 54:361.
MEWALDT, L.R. 1956. Nesting behavior of the Clark Nutcracker. Condor, 58:3-23.
MEZHENNY, A.A. 1961. Food competitors, enemies and diseases. *In* Egorov, O.V. (ed.), Ecology and Economics of the Yakut Squirrel. Moscow: Akademiya Nauk. (In Russian).
MEZHENNY, A.A. 1964. Biology of the nutcracker (*Nucifraga caryocatactes macrorhynchus*) in south Yakutia. Zoologischeskii Zhurnal, 43:1679-1687. (In Russian).
MIROV, N.T. 1967. The Genus *Pinus*. New York: The Ronald Press Co. 602 pp.
PALLAS, P.S. 1811. Zoographia Rosso-Asiatica, Vol. 1.
PARKES, K.C. 1953. The incubation patch in males of the suborder Tyranni. Condor, 55:218-219.
PETERSON, R.T. 1969. A Field Guide to Western Birds. Boston: Houghton-Mifflin Co. 366 pp.
PILGER, R. 1926. Genus *Pinus*. *In* Engler, A., and K. Prantl (eds.), Die natürlichen Pflanzenfamilien, Vol. 13, Gymnospermae. Leipzig: Wilhelm Engelmann.
PORTENKO, L.A. 1948. Neck pouches in birds. Priroda, 37: 50-54. (In Russian).

REIMERS, N.F. 1953. The food of the nutcracker and its role in the dispersal of the Cedar-pine in the mountains of Khamar-Daban. Lesnoe Khozyaistvo, 1:63-64. (In Russian).

REIMERS, N.F. 1959. The nesting of the Long Billed Nutcracker in Central Siberia. Zoologischeskii Zhurnal, 38:907-915. (In Russian).

SHAW, G.R. 1914. The Genus *Pinus*. Arnold Arboretum Publication No. 5. Boston: Houghton Mifflin Co.

SHAW, G.R. 1924. Notes on the genus *Pinus*. Arnold Arboretum Journal, 5:225-227.

SMITH, C.C. 1968. The adaptive nature of social organization in the genus of three squirrels *Tamiasciurus*. Ecological Monographs, 38:31-63.

SMITH, C.C. 1970. The coevolution of pine squirrels (*Tamiasciurus*) and conifers. Ecological Monographs, 40: 349-371.

SMITH, C.C. 1975. The coevolution of plants and seed predators; pp. 53-77. *In* Gilbert, L.E., and P.H. Raven (eds.), Coevolution of Animals and Plants. Austin, Texas: University of Texas Press.

SMITH, C.C., and R.P. BALDA. 1979. Competition among insects, birds and mammals for conifer seeds. American Zoologist, 19:1065-1083.

SUDWORTH, G.B. 1908. Forest Trees of the Pacific Slope. U.S. Department of Agriculture.

SWANBERG, P.O. 1956. Incubation in the Thick-billed Nutcracker, *Nucifraga c. caryocatactes* (L.); pp. 278-297. *In* Wingstrand, K.G. (ed.), Bertil Hanström--Zoological Papers in Honour of His Sixty-fifth Birthday. Lund, Sweden.

SWANBERG, P.O. 1981. Kullstorleken hos nötkråka *Nucifraga caryocatactes* i Skandinavien, relaterad till foregående års hasselnöttillgång. Vår Fågelvärld, 40:399-408.

TEVIS, L., JR. 1953. Stomach contents of chipmunks and mantled squirrels in northeastern California. Journal of Mammalogy, 34:316-324.

TOMBACK, D.F. 1978. Foraging strategies of Clark's Nutcracker. Living Bird, 16:123-161.

TOMBACK, D.F. 1980. How nutcrackers find their seed stores. Condor, 82:10-19.

TOMBACK, D.F. 1981. Notes on cones and vertebrate-mediated seed dispersal of *Pinus albicaulis* (Pinaceae). Madroño, 28:91-94.

TOMBACK, D.F. 1982. Dispersal of whitebark pine seeds by Clark's Nutcracker: A mutualism hypothesis. Journal of Animal Ecology, 51:451-467.

TOMBACK, D.F., and K.A. KRAMER. 1980. Limber pine seed harvest by Clark's Nutcracker in the Sierra Nevada: Timing and foraging behavior. Condor, 82:467-468.

TRANQUILLINI, W. 1979. Physiological Ecology of the Alpine Timberline. Berlin, Heidelberg: Springer-Verlag. 137 pp.

TURCEK, F.J. 1967. Ökologische Beziehungen der Säugetiere und Gehölze. Slovenska Akademia Vied, Bratislava, 1:210.

TURCEK, F.J., and L. KELSO. 1968. Ecological aspects of food transportation and storage in the Corvidae. Communications in Behavioral Biology, Part A, 1:277-297.

VANDER WALL, S.B. 1982. An experimental analysis of cache recovery in Clark's Nutcracker. Animal Behaviour, 30: 84-94.

VANDER WALL, S.B., and R.P. BALDA. 1977. Coadaptations of the Clark's Nutcracker and the piñon pine for efficient seed harvest and dispersal. Ecological Monographs, 47: 89-111.

VANDER WALL, S.B., and R.P. BALDA. 1981. Ecology and evolution of food-storage behavior in conifer-seed-caching Corvids. Zeitschrift für Tierpsychologie, 56:217-242.

WEAVER, T., and D. DALE. 1974. *Pinus albicaulis* in central Montana: Environment, vegetation and production. American Midland Naturalist, 92:222-230.

WEST, N.E. 1968. Rodent-influenced establishment of ponderosa pine and bitter-brush seedlings in central Oregon. Ecology, 49:1009-1011.

WOODMANSEE, R.G. 1977. Clusters of limber pine trees: A hypothesis of plant-animal coaction. Southwestern Naturalist, 21:511-517.

FAHRENHOLZ'S RULE AND RESOURCE TRACKING: A STUDY OF HOST-PARASITE COEVOLUTION

Robert M. Timm

Department of Zoology
Field Museum of Natural History
Chicago, Illinois

Fahrenholz's Rule and Resource Tracking are two hypotheses which describe host-parasite coevolution. Fahrenholz's Rule states that "In groups of permanent parasites the classification of the parasites usually corresponds directly [to] the natural relationships of the hosts," and Resource Tracking that "the parasite may track some particular and independently distributed resource on the host ...[where] we expect noncongruent host-parasite relationships."

As a test of these hypotheses, I examined approximately 20,000 chewing lice of the genus Geomydoecus *(Mallophaga: Trichodectidae) on pocket gophers of the genus* Geomys *(Rodentia: Geomyidae). Lice were obtained from all described subspecies of the* Geomys bursarius *complex and represent 590 individual hosts from 427 localities. In addition to qualitative features, 28 morphological characters were quantified for both adult male and female lice. The* Geomydoecus *proved to be quite variable geographically; however, there was little intrapopulation variability. This geographic variation is best represented taxonomically by recognizing 8 distinct monotypic species of lice, which cluster as 2 distinct groupings: a "northern" group, composed of* Geomydoecus geomydis, G. illinoensis, G. nebrathkensis, G. oklahomensis, *and* G. spickai *and a "southern" group, composed of* G. ewingi, G. heaneyi, *and* G. subgeomydis. *In no case was a single population of pocket gophers parasitized by more than one species of* Geomydoecus.

Because the pocket gophers are distributed allopatrically, and together with their chewing lice have limited dispersal ability, the Geomys-Geomydoecus *system is consistent with Fahrenholz's Rule. Bird ectoparasites have greater opportunity than mammal parasites to disperse to new species of hosts, and because of the complexity of feathers, have a greater range of niches available. Therefore, the ectoparasite fauna of birds is more diverse than that of mammals and that diversity is best explained by Resource Tracking. Coevolutionary relationships are greatly affected by the dispersal of parasites among host species. Fahrenholz's Rule and Resource Tracking are not conflicting hypotheses, but represent the ends of a continuum of dispersal opportunities and niche availability.*

Thus, the species of Mallophaga on Geomys *represent a lineage that has evolved in parallel with the pocket gophers (Fahrenholz's Rule), and the distribution and taxonomic relationships of the lice are a useful tool for elucidating the relationships of the pocket gophers. Based on this relationship several hypotheses are proposed concerning the relationships of the pocket gophers.*

INTRODUCTION

The hypothesis that the natural classification of certain groups of parasites parallels that of their hosts was first proposed by Fahrenholz in the late 1800's. Fahrenholz's conclusions on the phylogenetic parallelism of parasites and hosts were based on work on feather mites (Acarina), and later hypothesized that such a relationship held for the sucking lice (Anoplura) and chewing lice (Mallophaga). Eichler (1948, p. 599) subsequently coined the term "Fahrenholz's Rule" and defined this hypothesis as follows: "In groups of permanent parasites the classification of the parasites usually corresponds directly with the natural relationships of the hosts."

The basis of the hypothesis is the assumption that, at some point in the evolutionary history of host and parasite, the ancestral parasite enters a close association with the ancestral host, after which both evolve and speciate together. Thus, speciation and degree of divergence in the host taxa are parallel to those of their parasites. Although it has been suggested that the classification of various ectoparasites may be utilized as a taxonomic tool in the classification of their vertebrate hosts (Clay, 1970; Hopkins, 1949a, b; Rothschild and Clay, 1957), little work has actually been done along these lines (Timm, 1975).

Wenzel et al. (1966) and Machado-Allison (1967) drew similar conclusions concerning the relationships of certain phyllostomatid bats, based on parasite relationships. They suggested that the vampire bats, then recognized as a family, Desmodontidae, were most closely related to the phyllostomatid bats, and that the Chilonycterinae, then a subfamily of Phyllostomatidae, should be elevated to familial status. Subsequent systematic studies of the vampire bats confirmed this and reduced the desmodontids to a subfamily of the Phyllostomatidae (Forman et al., 1968), and elevated the Chilonycterinae to familial status as the Mormoopidae (Smith, 1972). Holland (1958; 1963) proposed two hypotheses concerning the taxonomy of arctic ground squirrels, *Spermophilus parryii*, based on the taxonomy and distribution of their parasitic fleas that can be explained by: 1) a close affinity between the New World and Old World arctic ground squirrels; and 2) a distinct arctic-subarctic division in the New World ground squirrels. Nadler and Hoffmann (1977) later concluded that northern and southern populations of New World arctic ground squirrels are more similar to the Siberian ground squirrels than either is to the other.

A second and contrasting model of host-parasite coevolution, is Resource Tracking. Here "...the parasite may track some particular and *independently* distributed resource on the host. Here we expect noncongruent host-parasite relationships" (Kethley and Johnston, 1975, p. 232). This hypothesis was based on a revision of the quill mites (Syringophilidae) of birds (Kethley, 1970). The Resource Tracking model predicts that there is no direct parallel relationship between the taxonomy of hosts and that of their parasites, but rather that the parasites are tracking a resource, such as a particular type of skin, hair, feathers, quill wall thickness, etc. on the host or hosts. Resource availability then is the limiting factor controlling the occurrence of parasites on various species of hosts. In support of the Resource Tracking theory, they demonstrated that taxonomy at the generic level of the syringophilid mites does not correlate directly with the taxonomy of their bird hosts; closely related genera of mites are found on birds of different orders. They concluded that the major patterns of syringophilid inter-relationships are independent of the major patterns of host inter-relationships. In a study of the alcid lice, Eveleigh and Amano (1977) found little correlation between host relationships and parasite relationships and concluded that Resource Tracking best described their observations.

While analyzing evolutionary patterns of parasites and their hosts, I became interested in the contradictions between Fahrenholz's Rule and Resource Tracking (Timm, 1979) and saw these apparent contradictions as central to our understanding of parasite evolution. It appeared that coevolutionary relationships can be greatly affected by the dispersal of parasites from one host species to another. The unstated assumption behind Fahrenholz's Rule is that there is no gene flow of parasites between unrelated hosts. The implicit assumption behind the Resource Tracking model is that it is equally likely for all species of parasites to disperse to any host, and that at least some of those dispersers will survive and reproduce on the new host. Examples chosen from highly mobile hosts with frequent interspecific contact like communal

---

FIGURE 1. An adult female plains pocket gopher, Geomys bursarius.

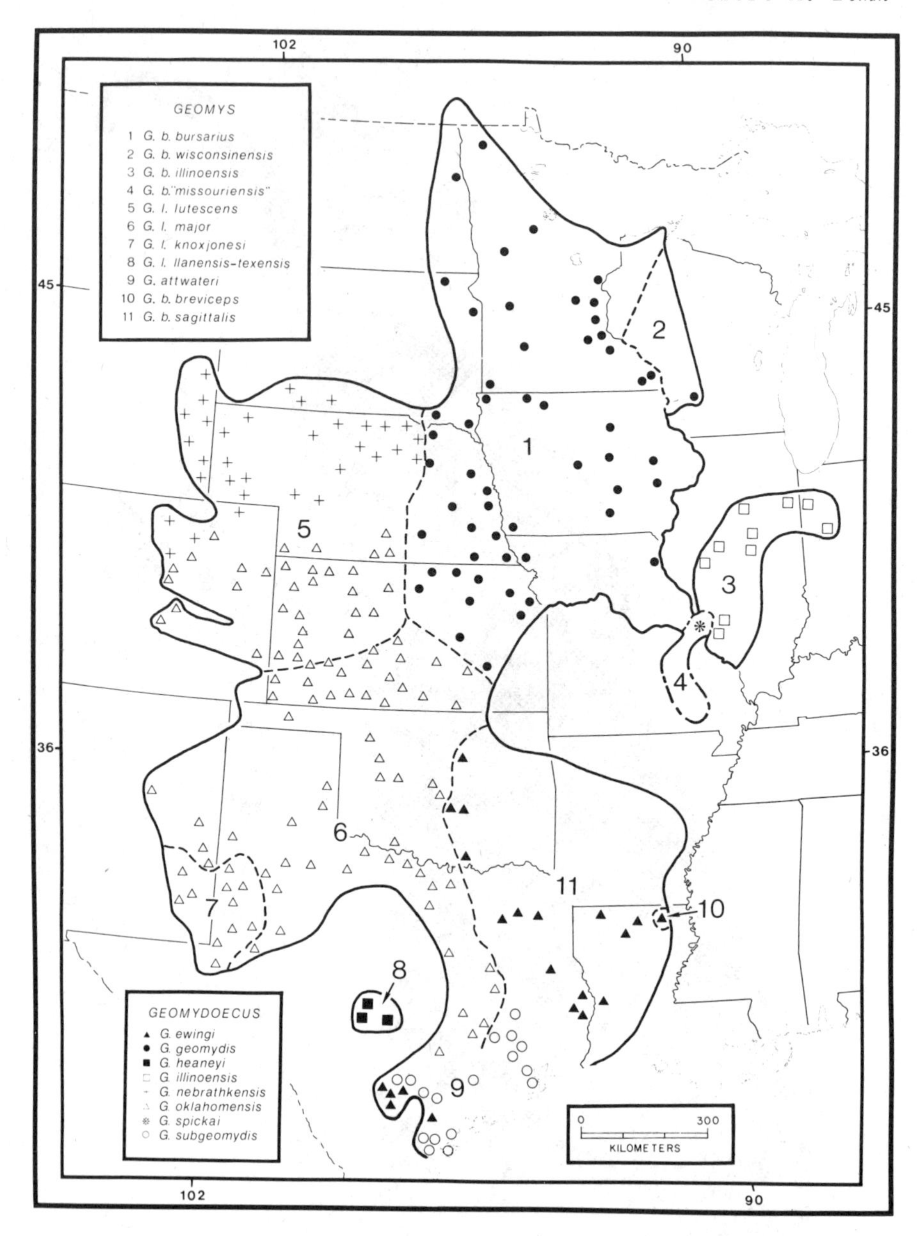
102
90
45
36
GEOMYS
1 G. b. bursarius
2 G. b. wisconsinensis
3 G. b. illinoensis
4 G. b."missouriensis"
5 G. l. lutescens
6 G. l. major
7 G. l. knoxjonesi
8 G. l. llanensis-texensis
9 G. attwateri
10 G. b. breviceps
11 G. b. sagittalis
GEOMYDOECUS
G. ewingi
G. geomydis
G. heaneyi
G. illinoensis
G. nebrathkensis
G. oklahomensis
G. spickai
G. subgeomydis
0 300
KILOMETERS

birds or bats, could be misleading, especially if different parasite species possess different dispersal abilities. Yet most examples used to support individual models are from the parasites of birds and bats. The dispersal potential of the parasites does not figure explicitly in either hypothesis of host-parasite coevolution.

To test the Fahrenholz's Rule and Resource Tracking hypotheses, I chose to examine the chewing lice of the genus *Geomydoecus* (Mallophaga: Trichodectidae) on pocket gophers of the genus *Geomys* (Rodentia: Geomyidae). All five genera of the family Geomyidae are parasitized by chewing lice of the genus *Geomydoecus*. Fifteen species of *Geomydoecus* are recognized from *Geomys* (Timm, 1979; Timm and Price, 1980).

Pocket gophers of the genus *Geomys* are found throughout much of the prairie region of central North America. *Geomys* ranges from Georgia and Florida west to New Mexico, and from extreme southern Manitoba south to Tamaulipas. Currently *Geomys* is divided into nine species with some 23 subspecies, but this classification is in a current state of flux and the status of several species and subspecies is uncertain (Heaney and Timm, 1983; Honeycutt and Schmidly, 1979). The most widely distributed species of *Geomys*, *G. bursarius* (fig. 1), is found throughout much of the midwestern United States. It ranges from Illinois and Indiana west to Colorado and New Mexico, and from extreme southern Manitoba to southern Texas (fig. 2). Regarded as a single species for the past 30 years, *Geomys bursarius* is composed of populations originally described as

---

FIGURE 2. Map of the distribution of the 8 species of Geomydoecus *that parasitize* Geomys attwateri, Geomys breviceps, Geomys bursarius, *and* Geomys lutescens. *The inner lines (dashes represent the boundaries between taxa of pocket gophers.*

four distinct species: *Geomys bursarius* (Shaw), *Geomys breviceps* Baird, *Geomys lutescens* Merriam, and *Geomys texensis* Merriam. Recent studies have shown that there is little or no gene flow between several of the supposed subspecies of *Geomys bursarius* (Bohlin and Zimmerman, 1982; Heaney and Timm, 1983). Thus, some populations of *Geomys bursarius* form genetically distinct populations; hence, in this paper I will refer to these pocket gophers as the *Geomys bursarius* complex.

The *Geomys-Geomydoecus* host-parasite system offers opportunities for investigating the Fahrenholz's Rule and Resource Tracking hypotheses because: 1) species of *Geomys* are distributed either allopatrically or parapatrically, no species are sympatric, and 2) both the pocket gophers and their lice have extremely limited dispersal ability. Thus, the patterns observed are quite likely the primary pattern of parasite-host coevolution and not a result of secondary or tertiary recolonization. Additionally, this host-parasite system is unique in that the morphological species concept utilized for the classification of the lice was tested in the field and found to represent reproductively isolated populations (Timm, unpubl.).

METHODS

Lice were obtained from all species and subspecies of *Geomys* currently recognized, as well as from nine no longer recognized subspecies. Large samples of lice were obtained whenever possible from numerous localities throughout the range of each subspecies of pocket gopher, and are deposited in the entomology collection of the University of Minnesota, St. Paul. Pocket gophers are deposited in the Bell Museum of Natural History at the University of Minnesota, Field Museum of Natural History, Museum of Natural History at the University of Kansas, and the Museum of Zoology, University of Michigan.

Qualitative features and 28 morphological characters were quantified for adult male and female lice. Three BMDP programs were utilized for the multivariate statistical analysis: principal components analysis (BMDP4M), discriminant function analysis (BMDP7M), and cluster analysis (BMDP2M) (Timm and Price, 1980). All taxonomic decisions concerning the lice were made independently of the taxonomy and distributions of the pocket gophers; for details concerning the taxonomic revision of the pocket gophers, see Heaney and Timm (1983). Approximately 20,000 lice from some 600 individual hosts of the *Geomys bursarius* complex, representing 427 separate localities have been examined. The abundance of this material has permitted a thorough revision of the *Geomydoecus* on the *Geomys bursarius* complex and resulted in redescription of the four previously recognized species of lice, description of four additional species, and refinement of our knowledge of the distribution of lice on pocket gopher taxa (Timm, 1979; Timm and Price, 1980).

EVOLUTION AND NATURAL HISTORY OF POCKET GOPHER LICE

Evolution

Osborn (1891) was the first to mention finding lice on pocket gophers; he described *Trichodectes geomydis* on the basis of several specimens off the plains pocket gopher, *Geomys bursarius* (Shaw), from Ames, Iowa. In the next 30 years, four additional species of lice, *Trichodectes californicus* Chapman, *T. expansus* Duges, *T. scleritus* McGregor, and *T. thomomys* McGregor were described. Later, Ewing (1929) described a new genus, *Geomydoecus*, within the family Trichodectidae for this group of lice.

In 1897, Chapman described *Trichodectes californicus* on

the basis of a single female obtained from a pocket mouse, *Perognathus* sp. The erroneous designation of *Perognathus* as the host for *Geomydoecus californicus* caused a good deal of confusion because the family Geomyidae, which includes all the pocket gophers, and the family Heteromyidae, which includes the genus *Perognathus*, are closely related, leading one to expect to find closely related lice on the two families (i.e. see Jellison, 1942; Paine, 1912). However, the heteromyid rodents are parasitized by lice of the order Anoplura (genus *Fahrenholzia*), whereas the geomyids are parasitized by lice of the order Mallophaga (genus *Geomydoecus*). Werneck (1945) obtained numerous individuals of *Geomydoecus californicus* from Botta's pocket gopher, *Thomomys bottae bottae* (Eydoux and Gervais) and designated that gopher as the type host of *G. californicus*. It is now well established that all members of the genus *Geomydoecus* are obligatory ectoparasites of pocket gophers. There is no evidence to indicate that *Geomydoecus* can reproduce on hosts other than pocket gophers.

The family Geomyidae is a strictly New World family of the rodent suborder Sciuromorpha. There are five genera, found across most of the western two-thirds of North America, from central Canada south to northern Colombia. All genera of pocket gophers are parasitized by one of more species of *Geomydoecus*. The family Geomyidae is the only family of the suborder Sciuromorpha parasitized by Mallophaga; all other sciuromorphs are parasitized by anoplurans.

Currently, 102 specific and subspecific taxa are recognized in the genus *Geomydoecus*. These are morphologically divided into two distinct subgenera (see Price and Emerson, 1972). The nominate subgenus is found on all five genera of pocket gophers, whereas, the subgenus *Thomomydoecus* is found only on pocket gophers of the genus *Thomomys*.

It seems likely that the host-parasite association between *Geomydoecus* and the geomyids began at least as early as the late Miocene or early Pliocene if Russell's (1968) phyletic tree of the geomyids is correct. Geomyids probably had their origin in the southwestern United States and northern Mexico, then radiated out in all directions. Russell (1968) postulated that *Thomomys*, the most divergent genus of geomyids, split off from the main line of geomyids in the early Pliocene. It seems likely then, that this early Pliocene split in the hosts resulted in the two distinct subgenera of *Geomydoecus* we see today. However, all *Thomomys* are also parasitized by members of the subgenus *Geomydoecus* in addition to some having *Thomomydoecus*. This probably is a result of a secondary reinfestation, as these lice (subgenus *Geomydoecus*) on *Thomomys* are very similar to those on the genera *Geomys* and *Pappogeomys*. *Geomydoecus* is similar morphologically to two widespread genera of carnivore lice, *Trichodectes* and *Neotrichodectes*. Trichodectid lice are common and widespread on carnivores, but are not found on sciuromorph rodents other than geomyids. Therefore, it seems probable that the ancestral geomyids were infested with trichodectid lice from an ancestral carnivore.

Natural History

No individual or population of *Geomys* has ever been found to be parasitized by more than one species of *Geomydoecus*. The population of lice on an adult *Geomys* varies seasonally, but averages over 500 individuals during the summer months, with some individual gophers having as many as 2,000 lice. There is usually a one to one sex ratio (tables 1 and 2), but two parthenogenetic species have been found (Price and Timm, 1979). All stages of lice, including eggs, are most abundant

Table 1. Species of Geomydoecus *in which the sex ratio of adults was significantly different than 1 to 1.*

Species	Ratio F:M	N	Significance X^2	Significance p	Source of Data
birneyi	1.6:1	253	5.98	$p<.05$	Price & Hellenthal, 1980c
crovelloi	.7:1	647	8.20	$p<.005$	Price & Hellenthal, 1981a
expansus	.9:1	2,907	5.83	$p<.05$	Price & Hellenthal, 1975a
idahoensis	1.3:1	416	9.74	$p<.005$	Price & Hellenthal, 1980a
mobilensis	191:0*	191	-	-	Price, 1975
musculi	1.7:1	412	12.63	$p<.005$	Price & Hellenthal, 1981b
nayaritensis	2.5:1	97	8.66	$p<.005$	Price & Hellenthal, 1981b
scleritus	500:1	1,000	500.00	$p<.001$	Price & Timm, 1979
tamaulipensis	20:1	42	17.19	$p<.001$	Price & Hellenthal, 1975b
thomomyus	2.1:1	308	18.94	$p<.001$	Price & Emerson, 1971

* *Males of* G. mobilensis *are unknown.*

Table 2. Species of Geomydoecus *with a 1 to 1 sex ratio. Data from Price and Emerson, 1971; Price and Hellenthal, 1975a, b, 1976, 1979, 1980a, b, c, d, 1981a, b; Price and Timm, 1979; Spicka, 1981; and Timm and Price, 1980.*

actuosi *N = 897*
albati *N = 414*
angularis *N = 192*
asymmetricus *N = 52*
aurei aurei *N = 543*
aurei grahamensis *N = 607*
bajaiensis *N = 204*
californicus *N = 1,309*
centralis *N = 3,574*
chapini *N = 38*
chihuahuae *N = 112*
clausonae *N = 504*
cliftoni *N = 65*
copei *N = 113*
coronadoi *N = 40*
costaricensis *N = 38*
dakotensis *N = 81*
dickermani *N = 127*
expansus *N = 2,907*
extimi *N = 88*
fulvi *N = 193*
genowaysi *N = 855*
geomydis *N = 984*
guadalupensis *N = 166*
heaneyi *N = 173*
hoffmanni *N = 101*
hueyi *N = 387*
illinoensis *N = 4,300*
jaliscoensis *N = 100*
johnhafneri *N = 112*
limitaris halli *N = 116*
limitaris limitaris *N = 561*
limitaris tolteci *N = 362*
martini *N = 606*
mcgregori *N = 137*
merriami *N = 49*
mexicanus *N = 40*
minor *N = 2,797*
neocopei *N = 36*
oklahomensis *N = 601*
oregonus *N = 84*
orizabae *N = 53*
panamensis *N = 290*
pattoni *N = 20*
polydentatus *N = 43*
shastensis *N = 456*
spickai *N = 4,835*
subcalifornicus *N = 4,755*
subnubili *N = 872*
texanus tropicalis *N = 61*
timmi *N = 330*
tolucae *N = 124*
traubi *N = 58*
truncatus *N = 30*
umbrini *N = 310*
ustulati clarkii *N = 219*
ustulati ustulati *N = 532*
wardi *N = 75*
warmanae *N = 2,073*
welleri welleri *N = 2,016*
welleri multilineatus *N = 1,013*
williamsi *N = 32*
yucatanensis *N = 101*
zacatecae *N = 686*

on the back of the head and nape of the neck, presumably because these areas are the most difficult for the gopher to reach while grooming. Grooming by the host probably is the main factor controlling densities of louse populations. The lice probably feed on scrapings of skin and hair from their hosts. Each individual egg, called a nit, is glued to a single hair; eggs hatch in approximately 10 days, with three nymphal stages lasting about 10 days each (Rust, 1974). Rust (1974) reported that *Geomydoecus oregonus* on *Thomomys bottae* in the Sacramento Valley of California reproduced throughout the year, as did their hosts, and that adult lice lived for 30 days. Price and Timm (1979) reported that *Geomydoecus scleritus*, a parthenogenetic species on pocket gophers in the southeastern United States reproduced throughout the year, as did the host population of pocket gophers. On northern *Geomys*, lice reproduce only during the spring, summer, and early fall, and the adults live several months. The northern populations of pocket gophers reproduce only during the spring and summer months. It seems likely that these lice are cueing in on the reproductive cycle of their hosts. The only ecto-parasites reproducing synchronously with their hosts are species that feed directly on blood (see Foster, 1969; Rothschild and Ford, 1964, 1966, 1969), where the reproductive steroids of the host presumably trigger the reproductive steroids of the parasite. However, pocket gopher lice feed

---

FIGURE 3. Pocket gopher louse, Geomydoecus geomydis. *A. Dorsal view of an adult female; B. dorsal view of an adult male; C. ventral view of an adult female; D. ventral view of a second instar. All specimens from* Geomys bursarius bursarius, *Washington County, Kansas. Length of A and C 1.3 mm, B 1.4 mm, and D 0.9 mm.*

A

B

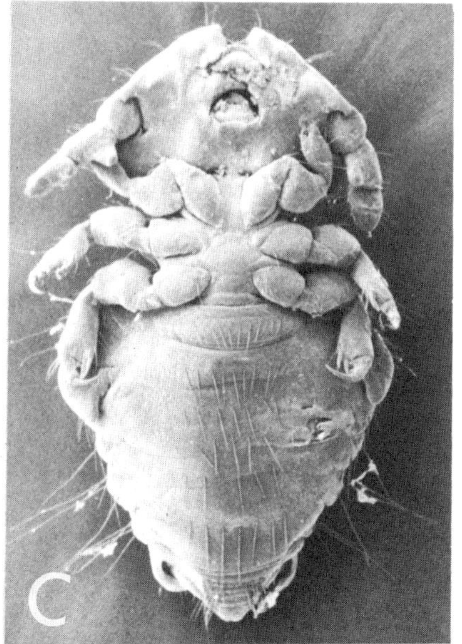
C

D

on dead tissue and not on blood. Hence, how they are cueing in on the reproductive cycle of the host is an interesting question.

Transmission of lice from one host to another occurs only upon direct contact between hosts, because lice, unlike other ectoparasites such as fleas and ticks, cannot live independently of the host. Because pocket gophers are fossorial and solitary, transmission may occur only during breeding or from a female to her offspring. I have found lice on 14 day old gophers, the approximate age when pocket gophers begin to develop a full pelage. A recently-dispersed young pocket gopher (*Geomys bursarius wisconsinensis*) which I captured had a population of over 350 lice, (*Geomydoecus geomydis* see figs. 3 and 4), including all stages of the life cycle; apparently dispersing gophers carry with them a founder population of lice.

There are few reports of *Geomydoecus* on hosts other than pocket gophers (see Timm and Price, 1980, for a review). These are presumed to be cases of stragglers or contamination because all have been single individuals and not populations. The five records of *Geomydoecus* found on long-tailed weasels, *Mustela frenata*, probably represent natural stragglers (see Timm and Price, 1980). Long-tailed weasels are major predators on pocket gophers, and probably picked up the lice from their prey.

Taxa of *Geomydoecus* are distinguished by morphology of the genitalia for both sexes, by differences in chaetotaxy,

---

FIGURE 4. Enlarged dorsal (A) and ventral (B) views of the head of an adult female Geomydoecus geomydis; *width of the head 0.46 mm. Locality and host data same as in fig. 3.*

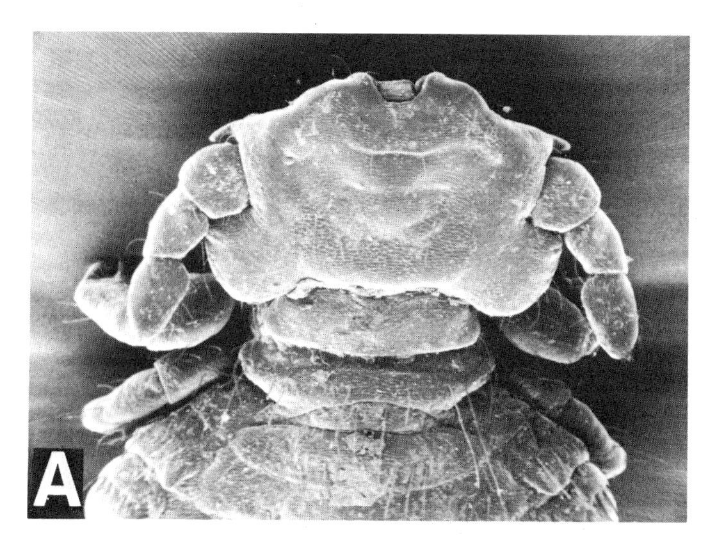
A

B

size, and by the distinctive antennal scape of the males. Thus, it is likely that the morphospecies are either a subset or an aggregate of the biological species. As a test of interbreeding (or lack of it) between morphospecies of *Geomydoecus*, the systematics of lice on populations of hybridizing pocket gophers was investigated. Data from two zones of hybridization in which the two parental populations of pocket gophers each had a different species of louse were obtained. In no cases were hybrid lice found; the morphospecies of *Geomydoecus* appear to be either true host-specific biological species, or aggregates of a biological species.

DISTRIBUTION OF LICE

The *Geomydoecus* on the *Geomys bursarius* complex cluster into two main groupings, the "northern group" and the "southern group" (Timm and Price, 1980). The "northern group" is composed of the "*geomydis*" complex and the "*oklahomensis*" complex of species; the "southern group" is composed of three species (see fig. 5).

Northern Group

I. Geomydis complex.

A) Geomydoecus geomydis--This louse is found on two subspecies of pocket gophers, *Geomys bursarius bursarius* (including *majusculus*) and *Geomys bursarius wisconsinensis*.

Heaney and Timm (1983) found a continuous clinal pattern of variation in both size and cranial characteristics (fig. 6) between the northern-most populations of *Geomys bursarius bursarius* and the southern-most populations described as *Geomys bursarius majusculus* in specimens of *Geomys* from Iowa, Kansas, Minnesota, Missouri, and Nebraska. We therefore

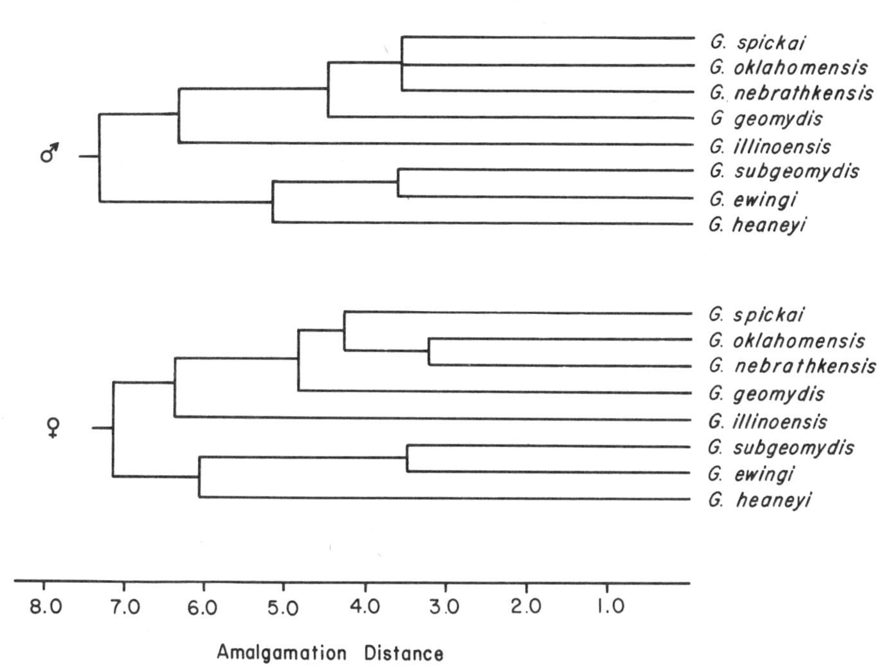

FIGURE 5. Distance phenogram of the cluster analysis of the 8 species of Geomydoecus. *(From Timm and Price, 1980. Reprinted, with permission, from the Journal of Medical Entomology).*

concluded that *Geomys bursarius majusculus* does not merit subspecific distinction. *Geomys bursarius wisconsinensis* (including a series from the type locality at Lone Rock, Richland Co., Wisconsin) is similar to *Geomys bursarius bursarius*, except for minor cranial characters mentioned by Jackson (1957) in his subspecific description of *Geomys bursarius wisconsinensis*. *Geomys bursarius bursarius* and *Geomys bursarius wisconsinensis* are chromosomally indistinguishable, each having 2N = 72, FN = 72, and X as a large acrocentric chromosome (Hart, 1978). The karyotype of *majusculus* from eastern Kansas is similar except that FN is 70. These subspecies of pocket gophers occupy the northeastern third of the range of *Geomys bursarius*. They are found in the

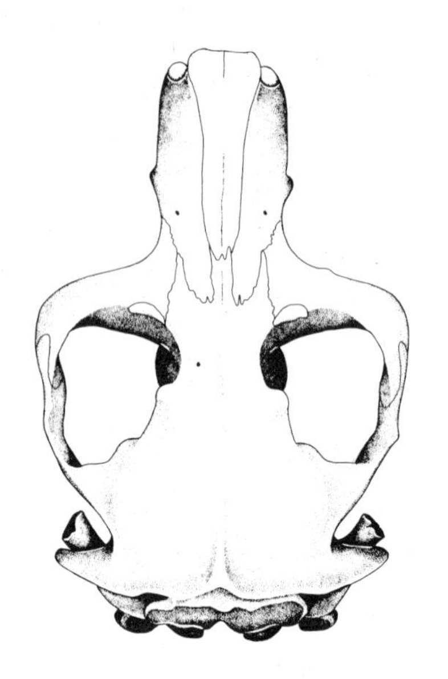

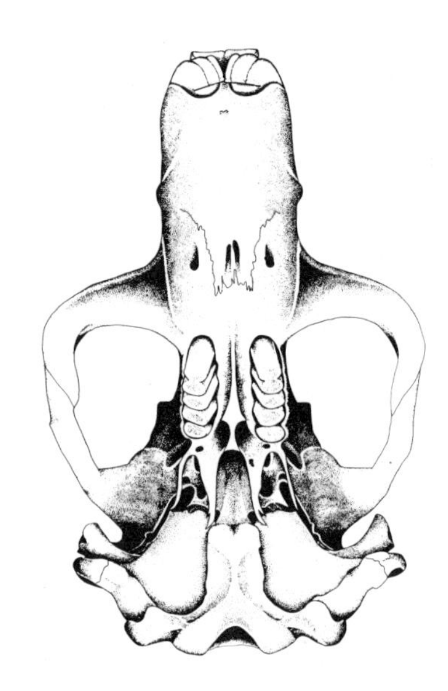

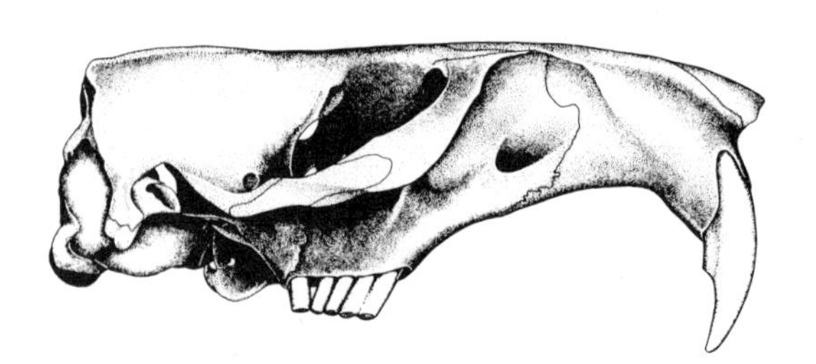

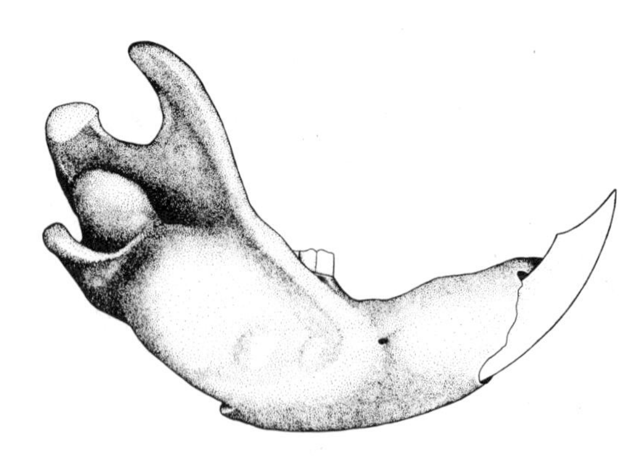

Bluestem prairie (*Andropogon-Panicum-Sorghastrum*), Oak savanna (*Quercus-Andropogon*), and the Bluestem-Oak-Hickory savanna (*Andropogon-Quercus-Carya*) vegetational communities (see Küchler, 1964). Undoubtedly these pocket gophers are a closely related group.

B) Geomydoecus illinoensis--This species is restricted to one subspecies of pocket gopher, *Geomys bursarius illinoensis*. Although definitely a member of the "*geomydis*" complex, this louse is quite distinct from *Geomydoecus geomydis*, suggesting that the two populations have been separated for a considerable period of time. Its host, *Geomys bursarius illinoensis*, is the northeastern-most subspecies of *Geomys* and is unique among *Geomys* in that nearly all specimens are black.

Hart (1978) found that *illinoensis* and *majusculus* have identical karyotypes, each having 2N = 72, FN = 70, and all chromosomes, including the X, are large acrocentrics. The karyotypes of *Geomys bursarius bursarius* and *Geomys bursarius wisconsinensis* are similar, but have an FN of 72 rather than 70. Cranial morphometrics suggest that *Geomys bursarius illinoensis* is a distinct subspecies most like the *Geomys bursarius bursarius* and *Geomys bursarius wisconsinensis* group, and that these two groups are distinct from the western and southern *Geomys* (Heaney and Timm, 1983). *Geomys bursarius illinoensis* occurs throughout central Illinois and extreme northwestern Indiana in the Bluestem Prairie (*Andropogon-Panicum-Sorghastrum*) - Oak-Hickory Savannah (*Quercus-Carya*) vegetational community, and is isolated geographically from

---

FIGURE 6. Cranium and mandible of an adult male plains pocket gopher, Geomys bursarius, *from Oakdale, Antelope County, Nebraska (FMNH 123429). Length of skull 59.2 mm, of mandible 43.9 mm.*

the more western subspecies by major river systems, the Mississippi and Illinois rivers on the west and the Kankakee River on the north.

II. Oklahomensis complex.

A) Geomydoecus oklahomensis--This louse is found on *Geomys lutescens knoxjonesi*, *Geomys lutescens lutescens* (in part), and *Geomys lutescens major* (including *industrius* and *jugossicularis*). The pocket gophers from southwestern Nebraska, western Kansas, Oklahoma, and Texas, and eastern Colorado and New Mexico are all parasitized by this species. Although this louse is the most variable and the most widely distributed of the eight species found on the *Geomys bursarius* complex, there is no evidence that any of the populations warranted classification as a distinct taxon. Principal components analysis and discriminant function analysis suggest that the lice on two populations of pocket gophers, described as *industrius* and *jugossicularis*, are a single population. Geographic variation between populations of *Geomydoecus* is evident between all other subspecies of the *Geomys bursarius* complex.

These pocket gophers have had a varied taxonomic history (Merriam, 1890, 1895; Villa and Hall, 1947; Russell and Jones, 1956; Jones, 1964; and Heaney and Timm, 1983). *Geomys lutescens* was first described from the Sand Hills of Nebraska as a subspecies of *Geomys bursarius*, but later elevated to full specific rank by Merriam. Merriam's species *lutescens* included the pocket gophers now considered *Geomys lutescens knoxjonesi*, *Geomys lutescens lutescens*, and *Geomys lutescens major*; and five other populations that are no longer considered valid subspecies, *Geomys lutescens hylaeus* Blossom, *Geomys* [*bursarius*] *industrius* Villa and Hall, *Geomys lutescens jugossicularis*

Hooper, *Geomys lutescens levisagittalis* Swenk, and *Geomys lutescens vinaceus* Swenk. Villa and Hall (1947) reduced *lutescens* to a subspecies of *Geomys bursarius* on the basis of supposed intergradation between the two populations; however, there is no intergradation between these populations (Heaney and Timm, 1983). Karyotypic variation in several populations of *Geomys* from western Texas and eastern New Mexico has been partially described (see Baker et al., 1973; Baker and Genoways, 1975; Hart, 1978), resulting in the description of a new subspecies, *Geomys* [*bursarius*] *knoxjonesi*. However, karyotypic variation in the more northern populations remains poorly understood (see Hart, 1978; Heaney and Timm, 1983; Timm et al., in press). In a recent revision, Heaney and Timm (1983) concluded that: 1) there is no justification for recognizing *industrius* and *jugossicularis* as subspecies distinct from *major*; 2) that the group of pocket gophers represented by *lutescens* and *major* is composed of closely related taxa; 3) there is little or no gene flow between *lutescens-major* and the group including *bursarius*, *illinoensis*, and *wisconsinensis*; and 4) these two groups are two distinct species.

The suggestion made by Baker and Genoways (1975) that *Geomys* [*bursarius*] *knoxjonesi* is most closely related to the *llanensis-texensis* group is unlikely for two reasons. First, their phenogram and two-dimensional projection of the cranial morphology of the gophers indicates that *knoxjonesi* is morphologically close to the *Geomys lutescens major* group. Second, *Geomys lutescens knoxjonesi* is parasitized by *Geomydoecus oklahomensis*, a member of the "northern" species group (Timm and Price, 1980) which is also found on *Geomys lutescens major* and *Geomys lutescens lutescens*. The populations of lice on *Geomys lutescens knoxjonesi* show some morphological differentiation from the main body of *Geomydoecus oklahomensis*, but this difference is not sufficient to warrant taxonomic distinction (Timm and Price, 1980). However, *Geomys lutescens llanensis* and *Geomys lutescens texensis* have a distinctive species of louse, *Geomydoecus heaneyi*, that is a member of the "southern" species group. *Geomydoecus oklahomensis* is restricted to those taxa of *Geomys* that are found in the short-grass and mixed grass prairies, which include the Bluestem grama prairie (*Andropogon-Bouteloua*) and Grama-Buffalo grass prairie (*Bouteloua-Buchloe*). These gophers previously included in Merriam's species *lutescens* are parasitized by closely related members of the "*oklahomensis*" complex (*Geomydoecus oklahomensis* and *G. nebrathkensis*) and all of those gophers occurring south of the Platte River are parasitized by *Geomydoecus oklahomensis* proper.

B) Geomydoecus nebrathkensis--Although this species is found only on one subspecies of pocket gopher, *Geomys lutescens lutescens*, it is not found throughout its range; it occurs on the pocket gophers north of the Platte River in northern Nebraska, northeastern Colorado, eastern Wyoming, and southern South Dakota.

Hart (1978) reported that *Geomys [bursarius] lutescens* was the most variable of the subspecies of *Geomys [bursarius]* that he karyotyped, with the FN ranging from 70-98. He did find, however, that the karyotype of a single female from Kansas (south of the Platte River) was identical to those of *industrius* and *major*, while those from north of the Platte River were quite different, having numerous biarmed autosomes and a large metacentric X chromosome. On the basis of cranial morphology, Heaney and Timm (1983) agreed that *Geomys lutescens lutescens* was extremely variable geographically, but concluded that there was inadequate justification for splitting *lutescens* until their cytogenetics are better understood. *Geomydoecus nebrathkensis* occurs on the *Geomys lutescens lutescens* north of the Platte River in the Nebraska Sandhills prairie (*Andropogon-Calamovilfa*), Grama-buffalo grass (*Bouteloua-Buchloe*), and Wheatgrass-bluestem-needlegrass (*Andropyron-Andropogon-Stipa*) grassland communities of Küchler (1964).

C) Geomydoecus spickai--This louse is found on only one previously recognized subspecies of pocket gopher, *Geomys bursarius missouriensis*.

In his paper describing *Geomys bursarius missouriensis* as a distinct subspecies, McLaughlin (1958) stated that "The pocket gophers in Missouri represent a zoogeographical enigma ... The population of *G. b. missouriensis* is separated geographically from other populations of *G. bursarius* by a wide hiatus on the north, west and south. Only on the east does it approach the radically different *Geomys bursarius illinoensis*, which occupies the opposite bank of the Mississippi River..." Heaney and Timm (1983) found the taxa originally described as *Geomys bursarius missouriensis* to represent a composite of two different species, *Geomys bursarius* and *Geomys lutescens*. Those pocket gophers just south of the Missouri River were not significantly different from the pocket gophers to the north and west [*Geomys bursarius bursarius*], and the southern populations of *missouriensis* were identified as *Geomys lutescens*, although there is at least a 200 mile gap between these and the nearest populations of *Geomys lutescens* in northeastern Oklahoma. Additional specimens of both the pocket gophers and lice are needed to clarify this problem. Pocket gophers in southern Missouri apparently are restricted to "islands" of prairie within the Oak-Hickory forest (*Quercus-Caraya*), but little is known concerning the present distribution and habitat requirements of these pocket gophers.

Southern Group

A) Geomydoecus heaneyi--This louse is found on only two subspecies of pocket gophers, *Geomys lutescens llanensis* and *Geomys lutescens texensis*. These two pocket gophers occur in the Central Basin, a restricted area of the Edwards Plateau of central Texas. The Central Basin is a relatively small region with sandy clay and sandy loam soils (Pedernales and Tishomingo soils, see Carter, 1931), which support a Mesquite-Oak savanna (*Prosopis-Quercus-Andropogon*) vegetational community. *Geomys lutescens texensis* is found in a limited area in Mason and McCulloch counties and *Geomys lutescens llanensis* is found just to the east in a restricted area of Llano and San Sabo counties (Dalquest and Kilpatrick, 1973). Pocket gophers in the Central Basin are isolated from other populations of *Geomys* by clay soils that are unsuitable for *Geomys* and by the Colorado River (Davis, 1940).

Historically, the taxonomy of these two pocket gophers could best be described as unstable, in part because of striking differences in cranial morphology despite close geographical proximity. *Texensis*, first described as a distinct species (Merriam, 1895), was reduced to a subspecies of *Geomys breviceps* by Davis (1938). *Llanensis* was first described as a subspecies of *Geomys breviceps* by Bailey (1905), who stated that "While closely resembling *texensis* externally ...it needs but a cursory examination of the skulls to show that this form has no connection with that species." Davis (1940) transferred *llanensis* to a subspecies of *Geomys lutescens*. Later, *lutescens* was assigned as a subspecies of *Geomys bursarius* (Villa and Hall, 1947). Baker (1950) found what he considered hybrids between *texensis* and *llanensis*, so transferred *texensis* from the *Geomys breviceps* group to *Geomys bursarius*. Hence, both *llanensis* and *texensis* are classified today as subspecies of *Geomys lutescens*. Hart (1978) examined the karyotypes of both subspecies and found them to be similar (2N of 70-72, FN of 68-70 and all chromosomes acrocentric). He suggested that this group was possibly derived from *major*. *Geomydoecus heaneyi* is a very distinctive louse and this implies, as does the distinctive cranial morphology and karyotype data, that these two subspecies of pocket gophers have been genetically isolated for some time from the neighboring populations of other pocket gophers.

B) Geomydoecus subgeomydis--This species of louse is found on two taxa of pocket gophers, *Geomys attwateri* and *Geomys breviceps sagittalis*.

The population of pocket gophers now referred to as *Geomys breviceps sagittalis* was originally described as six different subspecies (*sagittalis*, *brazensis*, *dutcheri*, *ludemani*, *pratincola*, and *terricolus*). As a result of a morphometric and chromosomal analysis, Honeycutt and Schmidly (1979) concluded that the gophers of extreme southeastern Texas were best represented by recognizing a single subspecies, *Geomys [bursarius] sagittalis*, with a second taxon, *attwateri*, found to the west.

C) Geomydoecus ewingi--This species of louse is found on three taxa of pocket gophers, *Geomys attwateri*, *Geomys breviceps breviceps*, and *Geomys breviceps sagittalis*. The presence of *Geomydoecus ewingi* on *Geomys breviceps breviceps* supports Honeycutt and Schmidly's (1979) conclusion that *breviceps* is most closely related to *sagittalis*.

The geographic distribution of *attwateri* and *sagittalis* is in close, but not perfect, agreement with the boundary between the two species of lice, *Geomydoecus subgeomydis* and *Geomydoecus ewingi* (Timm and Price, 1980). In general, *Geomydoecus ewingi* is found on the eastern subspecies, *Geomys breviceps breviceps* and *Geomys breviceps sagittalis*, and *Geomydoecus subgeomydis* on the western species, *Geomys attwateri*. However, along the Brazos River some populations of *Geomys* that Honeycutt and Schmidly (1979) referred to the eastern subspecies, *Geomys b. sagittalis*, are parasitized by the western louse, *Geomydoecus subgeomydis*. Also, there is a population of the eastern louse, *Geomydoecus ewingi*, in Atascosa, Bexar, Goliad, and Wilson counties, Texas, that is separated by a wide hiatus of *Geomydoecus subgeomydis* from the main body of *Geomydoecus ewingi*. An analysis of the lice showed that there was no justification for splitting the two separated populations of *Geomydoecus ewingi* into two or more taxa (Timm and Price, 1980). These discrepancies suggest that our understanding of the systematics and host relationships of the *Geomys* and *Geomydoecus* of southeastern Texas warrants further investigations.

DISCUSSION

Taxonomic Hypotheses

All too often, systematists, in inferring phylogeny, overemphasize a single source of data, whether morphological, karyological, or biochemical. This tendency to overvalue one's own data can lead to erroneous conclusions (Atchley, 1972). Parasites, like any other single "taxonomic character," also cannot be expected to be the "one ideal" character to help systematists in differentiation of host taxa; however, parasites together with other data, can resolve many taxonomic problems.

Parasite data, like other forms of data, have distinct strengths and weaknesses. Parasite data can contribute unique character sets in that they are, at one and the same time, independent and evolving genotypes that are coevolving with the host. When the parasites have evolved in parallel with the host, they are useful taxonomically; however, not all species of parasites necessarily evolved in parallel with their hosts. Although parasites can evolve in parallel with their hosts, this does not imply that the evolutionary rate (or rate of speciation) is identical. Regardless of differences in evolutionary rates, when parasites have evolved in parallel with their hosts, they provide an extremely valuable tool for taxonomists.

The systematic conclusions concerning pocket gophers of the *Geomys bursarius* complex based on the relationships of their lice can be used as a set of hypotheses, which are as follow:

Implications for Pocket Gopher Taxonomy

1) The subspecies of plains pocket gophers, *Geomys bursarius bursarius* (including *majusculus*) and *Geomys bursarius wisconsinensis* form a closely related group.

2) *Geomys bursarius illinoensis* has been isolated from all other populations of *Geomys* for a considerable time, but is most closely related to the *bursarius-wisconsinensis* group.

3) The populations of *knoxjonesi*, *lutescens*, and *major* (including *industrius* and *jugossicularis*) form a closely related group.

4) The subspecies *llanensis* and *texensis* are most closely related to each other and have been isolated for a considerable time from other populations of *Geomys*.

5) The *llanensis-texensis* group is most closely related to the southeastern Texas populations (*attwateri-sagittalis*).

6) *Geomys breviceps breviceps* is most closely related to the adjacent *Geomys breviceps sagittalis*.

7) There is a distinct division between the southeastern subspecies (*attwateri-breviceps-sagittalis*) and the northern group (*knoxjonesi-lutescens-major*).

8) The northern *Geomys lutescens lutescens* (north of the Platte River) are distinct from the populations of *Geomys lutescens lutescens* south of the Platte River.

Role of Dispersal in Fahrenholz's Rule and Resource Tracking

Students of Mallophaga have long been puzzled by the following facts:

1) Frequently an individual bird may be parasitized by several species of Mallophaga, whereas in mammals it is uncommon to find more than one species of chewing louse on an individual.

2) Closely related species of mammals almost always are parasitized by closely related species of Mallophaga (Fahrenholz's Rule), whereas in birds this is only the case sometimes; frequently there is no evidence of close phylogenetic parallelism between birds and their parasites.

This difference between mammals and birds in speciation of their parasites is seen not only in the Mallophaga, but also in the parasitic mites. It was this lack of phylogenetic parallelism in quill mites (Syringophilidae) that led Kethley and Johnston (1975) to propose the Resource Tracking hypothesis.

In attempting to test the Fahrenholz's Rule and Resource Tracking hypotheses, I saw parasite dispersal rates as the most important single unknown factor once thorough taxonomic studies of both the host and parasite had been conducted. The unstated, but underlying assumption, behind Fahrenholz's Rule model is that there is no dispersal of parasites between unrelated hosts (or that dispersers do not survive). The unstated assumption behind the Resource Tracking model is that there is some opportunity for each species of parasites to disperse to and colonize any given host.

Dispersal (or the lack of dispersal) of individuals between populations is a primary factor in the speciation process. Wright (1931) has calculated that a dispersal rate as low as one individual per year between populations is enough to prevent genetic divergence (with no selection). If selection pressures are similar, a dispersal rate much less than one per year would be sufficient to maintain the genetic coupling of the two populations, and prevent speciation of lice on different host species.

Because the two models are based on conflicting assumptions concerning dispersal, the important question becomes whether dispersal of parasites among birds is different from that among mammals? Although few quantitative data on dispersal of Mallophaga are available, we do have some anecdotal information that can be evaluated.

We can assume that dispersal of lice from a female host to her offspring is commonplace, and that parasites readily disperse between adults during copulation, communal roosting, huddling, etc. Thus, there is ample opportunity for dispersal and mixing of the gene pool of parasites between different members of a single host species. However, transfer of parasites between different species of hosts presents additional problems. First, lice are extremely specialized for an ectoparasitic mode of life. They are wingless with legs highly modified to cling to either feathers or to fur, but not to walk on other substrates. Second, lice can not live long once removed from their hosts; Askew (1971, p. 21) wrote that "Lice generally are very sensitive to temperature and most soon die when their host's body cools after death."

How then can Mallophaga disperse to other species of hosts? It has been known for more than a century that

Mallophaga are sometimes found attached to hippoboscid flies (Diptera: Hippoboscidae). There are some one hundred species of hippoboscids world-wide; most are parasitic on birds, although a few parasitize large mammals. In general, hippoboscid flies are not host-specific; a single species of hippoboscid can be found on numerous families of birds and even on birds of different orders. Hill (1962) has shown that several species in the genus *Ornithomyia* select for size classes of birds or for birds of given habitats (woodland birds, moorland birds, etc.) rather than for closely related hosts. Most species of hippoboscids have wings and can fly readily. The lice cling to a fly by grasping a leg or wing with their mandibles; they are transported by the fly to a new host. This mode of transportation, termed phoresy, was defined by Farish and Axtell (1971, p. 17) as "a phenomenon in which one animal actively seeks out and attaches to the outer surface of another animal [phoriant] for a limited time during which the attached animal (termed the phoretic) ceases both feeding and ontogenesis, such attachment presumably resulting in dispersal from areas unsuited for further development, either of the individual or its progeny."

Phoresy by Mallophaga on hippoboscids is considered accidental or rare (Ansari, 1947). In a review of phoresy by Mallophaga, Clay and Meinertzhagen (1943) reported 13 new cases of lice being attached to some 200-300 hippoboscids they examined, and commented that these were "meager results." However, this magnitude of dispersal (0.5% of the flies they examined) is high when considered over evolutionary time. In recent reviews, Keirans (1975a, b) summarized records of 416 cases of phoresy, but felt that the only survival value for the lice involved would be if the hippoboscid fly transported it to another member of its host species. It has been noted that, when hippoboscid flies are carrying Mallophaga, frequently they have more than one (44% had two or more) (Clay and Meinertzhagen, 1943), and one fly was recorded with 31 lice (Peters, 1935). Clay (1949) suggested that phoresy on hippoboscids may have played an important role in the speciation of Mallophaga. It seems likely that phoresy on hippoboscid flies could have been a major factor in interspecific transfer of Mallophaga among birds. An interesting phoretic relationship has been suggested between bat fleas (*Lagaropsylla turba*) and ectoparasitic earwigs (*Arixenia esau*), both host-specific to the naked bat (*Cheiromeles torquatus*) (Marshall, 1977a,b; 1982). The eggs and larval stage of the flea are found in the bat guano on the floor of caves. Bat fleas are poor jumpers, and it has been suggested (Hutson, 1981) that the adult fleas reach the bats on the ceiling of the cave via transport by earwigs. Marshall (1977) found that 69% of the earwigs examined (201) carried fleas, with up to 40 fleas found on individual earwigs.

The feathers may be a second mechanism for interspecific transfer of bird ectoparasites. A considerable number of species of birds incorporate feathers of other birds into their own nests. This is especially true for the passerines. Bird feathers are also frequently incorporated into the nests of mammals, especially rodents. Thus, a founder population of a single female or even an unhatched louse egg on a loose feather may be deposited in the nest of a foreign host. If it is able to feed and reproduce on the new host, a successful transfer will have occurred.

A third mechanism for interspecific transfer of Mallophaga is communal use of dust baths (Clay, 1949; Hopkins, 1949b; Hoyle, 1938), but this mechanism is untested.

A fourth mechanism for interspecific transfer of bird ectoparasites may be in multiple use of holes for nesting. Cavities are a limited and highly prized resource for nesting.

A given cavity will host a successive array of species of birds throughout the nesting season and some birds will even expropriate other bird's nests (Welty, 1982). A parasite in the nest would find its host replaced by another species.

The opportunity for exchange of lice between different species of birds clearly exists. It is likely that the vast majority of interspecific exchanges do not survive, but because a single gravid (or parthenogenetic) female louse may give rise to populations on a new host, it seems likely that this sort of dispersal has produced the distribution patterns described as Resource Tracking. It is apparent that, in general, bird ectoparasites have more opportunities to reach a foreign host than do some groups of mammal ectoparasites.

In addition to the differences in dispersal opportunities for bird and mammal ectoparasites, birds and mammals differ in the complexity of niches available to the ectoparasites. Mammalian hair is relatively simple and uniform to lice, whereas avian feathers are complex in structure and variable across the body. Thus, parasites have several niches open to them on a single bird. The specialized niche provided by down, contour feathers, filoplumes, or the inner pulp cavity of the shaft may be more similar as a niche between closely related species of birds than they are among each other on the same bird. Cornell and Washburn (1979, p. 257) stated that, "parasite species richness asymptotically approaches an upper limit established mainly by host 'island' size and that recent evolutionary age has an insignificant effect on the number of species which attack the host." Birds constitute a "larger" island for colonization by parasites because of their diversity of niches available. The habitat richness on birds thus supports species richness of ectoparasites.

Successful transfer of lice between different taxa of birds is a rare event, but the probability of transfer of lice between taxa of pocket gophers of the genus *Geomys* is an exceedingly rare occurrence. Thus, the species of Mallophaga on *Geomys* represent a lineage that has evolved in parallel to the pocket gophers (Fahrenholz's Rule), and the relationships of the lice may be an extremely useful tool in elucidating the relationships of the pocket gophers. Fahrenholz's Rule and Resource Tracking are not conflicting hypotheses. Rather, they apparently represent the ends of a continuum based on dispersal opportunities and niches available to the parasite.

ACKNOWLEDGMENTS

I thank Hugh H. Genoways, Robert S. Hoffmann, James L. Patton, David J. Schmidly, and Henry W. Setzer for allowing pocket gopher skins under their care to be brushed for lice; Elmer C. Birney, Barbara L. Clauson, Edwin F. Cook, Lawrence R. Heaney, L. Henry Kermott, John B. Kethley, Ke Chung Kim, Matthew H. Nitecki, Bruce D. Patterson, Roger D. Price, and Rupert L. Wenzel provided helpful discussion and comments on the manuscript, and Rosanne Miezio illustrated the pocket gopher and skull. Special thanks are to Barbara L. Clauson, Lawrence R. Heaney, and Roger D. Price for assistance on various aspects of this project.

LITERATURE CITED

ANSARI, M.A.R. 1947. Association between the Mallophaga and the Hippoboscidae infesting birds. Journal of the Bombay Natural History Society, 46:509-516.

ASKEW, R.R. 1971. Parasitic Insects. New York: American Elsevier Publishing Company, 316 pp.

ATCHLEY, W.R. 1972. The chromosome karyotype in estimation of lineage relationships. Systematic Zoology, 21:199-209.

BAILEY, V. 1905. Biological survey of Texas. North American Fauna, 25:1-222.

BAKER, R.H. 1950. The taxonomic status of *Geomys breviceps texensis* Merriam and *Geomys bursarius llanensis* Bailey. Journal of Mammalogy, 31:348-349.

BAKER, R.J., and H.H. GENOWAYS. 1975. A new subspecies of *Geomys bursarius* (Mammalia: Geomyidae) from Texas and New Mexico. Occasional Papers The Museum, Texas Tech University 29:1-18.

BAKER, R.J., S.L. WILLIAMS, and J.C. PATTON. 1973. Chromosomal variation in the plains pocket gopher, *Geomys bursarius major*. Journal of Mammalogy, 54:765-769.

BOHLIN, R.G., and E.G. ZIMMERMAN. 1982. Genic differentiation of two chromosome races of the *Geomys bursarius* complex. Journal of Mammalogy, 63:218-228.

CARTER, W.T. 1931. The soils of Texas. Texas Agricultural Experiment Station, Bulletin No. 431:1-192.

CHAPMAN, B. 1897. Two new species of Trichodectes (Mallophaga). Entomological News, 8:185-187.

CLAY, T. 1949. Some problems in the evolution of a group of ectoparasites. Evolution, 3:279-299.

CLAY, T. 1970. The Amblycera (Phthiraptera: Insecta). Bulletin of the British Museum (Natural History) Entomology, 25:73-98.

CLAY, T. and R. MEINERTZHAGEN. 1943. The relationship between Mallophaga and hippoboscid flies. Parasitology, 35:11-16.

CORNELL, H.V., and J.D. WASHBURN. 1979. Evolution of the richness-area correlation for cynipid gall wasps on oak trees: A comparison of two geographic areas. Evolution, 33:257-274.

DALQUEST, W.W., and W. KILPATRICK. 1973. Dynamics of pocket gopher distribution on the Edwards Plateau of Texas. Southwestern Naturalist, 18:1-9.

DAVIS, W.B. 1938. Critical notes on pocket gophers from Texas. Journal of Mammalogy, 19:488-490.

DAVIS, W.B. 1940. Distribution and variation of pocket gophers (genus *Geomys*) in the southwestern United States. Texas Agricultural Experiment Station, Bulletin No. 590: 1-38.

EICHLER, W. 1948. Some rules in ectoparasites. Annals and Magazine of Natural History, Series 12, Vol. 1:588-598.

EVELEIGH, E.S., and H. AMANO. 1977. A numerical taxonomic study of the mallophagan genera *Cummingsiella* (=*Quadraceps*), *Saemundssonia* (Ischnocera: Philopteridae), and *Austromenopon* (Amblycera: Menoponidae) from alcids (Aves: Charadriiformes) of the northwest Atlantic with reference to host-parasite relationships. Canadian Journal of Zoology, 55:1788-1801.

EWING, H.E. 1929. A manual of external parasites. Springfield Illinois: Charles C. Thomas, Publisher, 225 pp.

FARISH, D.J., and R.C. AXTELL. 1971. Phoresy redefined and

examined in *Macrocheles muscaedomesticae* (Acarina: Macrochelidae). Acarologia, 13:16-29.

FORMAN, G.L., R.J. BAKER, and J.D. GERBER. 1968. Comments on the systematic status of vampire bats (family Desmodontidae). Systematic Zoology, 17:417-425.

FOSTER, M.S. 1969. Synchronized life cycles in the orange-crowned warbler and its mallophagan parasites. Ecology, 50:315-323.

HALL, E.R. 1981. The mammals of North America. Vol. 1. New York: John Wiley & Sons, 600 pp.

HART, E.B. 1978. Karyology and evolution of the plains pocket gopher, *Geomys bursarius*. Occasional Papers, Museum of Natural History, University of Kansas, 71:1-20.

HEANEY, L.R., and R.M. TIMM. 1983. Relationships of pocket gophers of the genus *Geomys* from the central and northern Great Plains. Miscellaneous Publication, Museum of Natural History, University of Kansas. (In press)

HELLENTHAL, R.A., and R.D. PRICE. 1980. A review of the *Geomydoecus subcalifornicus* complex (Mallophaga: Trichodectidae) from *Thomomys* pocket gophers (Rodentia: Geomyidae), with a discussion of quantitative techniques and automated taxonomic procedures. Annals of the Entomological Society of America, 73:495-503.

HILL, D.S. 1962. A study of the distribution and host preferences of three species of *Ornithomyia* (Diptera: Hippoboscidae) in the British Isles. Proceedings of the Royal Entomological Society of London, A. 37:37-48.

HOLLAND, G.P. 1958. Distribution patterns of northern fleas (Siphonaptera). Proceedings of the Tenth International Congress of Entomology, 1:645-658.

HOLLAND, G.P. 1963. Faunal affinities of the fleas (Siphonaptera) of Alaska with an annotated list of species; pp. 45-63. *In* Gressitt, J.L. (ed.), Pacific basin biogeography. Tenth Pacific Science Congress, Bishop Museum, Honolulu, 563 pp.

HONEYCUTT, R.L., and D.J. SCHMIDLY. 1979. Chromosomal and morphological variation in the plains pocket gopher, *Geomys bursarius*, in Texas and adjacent states. Occasional Papers The Museum, Texas Tech University, 58:1-54.

HOPKINS, G.H.E. 1949a. The host-associations of the lice of mammals. Proceedings of the Zoological Society of London, 119:387-604.

HOPKINS, G.H.E. 1949b. Some factors which have modified the phylogenetic relationship between parasite and host in the Mallophaga. Proceedings of the Linnean Society of London, 161:37-39.

HOYLE, W.L. 1938. Transmission of poultry parasites by birds with special reference to the "English" or house sparrow and chickens. Transactions of the Kansas Academy of Science, 41:379-384.

HUTSON, A.M. 1981. Observations on host-finding by bat-fleas, with particular reference to *Ischnopsyllus simplex* (Siphonaptera; Ischnopsyllidae) in Great Britain. Journal of Zoology, 195:546-549.

JACKSON, H.H.T. 1957. An unrecognized pocket gopher from Wisconsin. Proceedings of the Biological Society of Washington, 70:33-34.

JELLISON, W.L. 1942. Host distribution of lice on native American rodents north of Mexico. Journal of Mammalogy, 23:245-250.

JONES, J.K., JR. 1964. Distribution and taxonomy of mammals of Nebraska. University of Kansas Publications, Museum of Natural History, 16:1-356.

KEIRANS, J.E. 1975a. A review of the phoretic relationship between Mallophaga (Phthiraptera: Insecta) and Hippoboscidae (Diptera: Insecta). Journal of Medical Entomology, 12: 71-76.

KEIRANS, J.E. 1975b. Records of phoretic attachment of Mallophaga (Insecta: Phthiraptera) on insects other than Hippoboscidae. Journal of Medical Entomology, 12:476.

KETHLEY, J.B. 1970. A revision of the family Syringophilidae (Prostigmata: Acarina). Contributions of the American Entomological Institute, 5:1-76.

KETHLEY, J.B., and D.E. JOHNSTON. 1975. Resource tracking patterns in bird and mammal ectoparasites. Miscellaneous Publications of the Entomological Society of America, 9: 231-236.

KÜCHLER, A.W. 1964. Potential natural vegetation of the conterminous United States. Special Publications, American Geographical Society No. 36:1-39 + map.

MACHADO-ALLISON, C.E. 1967. The systematic position of the bats *Desmodus* and *Chilonycteris*, based on host-parasite relationships (Mammalia; Chiroptera). Proceedings of the Biological Society of Washington, 80:223-226.

MARSHALL, A.G. 1977a. The earwigs (Insecta: Dermaptera) of Niah Caves, Sarawak. Sarawak Museum Journal, 25:205-209.

MARSHALL, A.G. 1977b. Interrelationships between *Arixenia esau* (Dermaptera) and molossid bats and their ecto-parasites in Malaya. Ecological Entomology, 2:285-291.

MARSHALL, A.G. 1982. The ecology of the bat ectoparasite *Eoctenes spasmae* (Hemiptera: Polyctenidae) in Malaysia. Biotropica, 14:50-55.

MCLAUGHLIN, C.A. 1958. A new race of the pocket gopher *Geomys bursarius* from Missouri. Contributions in Science, Los Angeles County Museum, 19:1-4.

MERRIAM, C.H. 1890. Descriptions of twenty-six new species of North American mammals. North American Fauna, 4:1-61.

MERRIAM, C.H. 1895. Revision of the pocket gophers, family Geomyidae, exclusive of the species of *Thomomys*. North American Fauna, 8:1-258.

NADLER, C.F., and R.S. HOFFMANN. 1977. Patterns of evolution and migration in the arctic ground squirrel, *Spermophilus parryii* (Richardson). Canadian Journal of Zoology, 55: 748-758.

OSBORN, H. 1891. The Pediculi and Mallophaga affecting man and the lower animals. United States Department of Agriculture, Division of Entomology, Bulletin 7:1-56.

PAINE, J.H. 1912. Notes on a miscellaneous collection of Mallophaga from mammals. Entomological News, 23:437-442 + plate.

PETERS, H.S. 1935. Mallophaga carried by hippoboscids. Annals of the Carnegie Museum, 24:57-58.

PRICE, R.D., and K.C. EMERSON. 1971. A revision of the genus *Geomydoecus* (Mallophaga: Trichodectidae) of the New World pocket gophers (Rodentia: Geomyidae). Journal of Medical Entomology, 8:228-257.

PRICE, R.D., and K.C. EMERSON. 1972. A new subgenus and three new species of *Geomydoecus* (Mallophaga: Trichodectidae) from *Thomomys* (Rodentia: Geomyidae). Journal of Medical Entomology, 9:463-467.

PRICE, R.D., and R.A. HELLENTHAL. 1975a. A reconsideration of *Geomydoecus expansus* (Duges) (Mallophaga: Trichodectidae) from the yellow-faced pocket gopher (Rodentia: Geomyidae). Journal of the Kansas Entomological Society, 48:33-42.

PRICE, R.D., and R.A. HELLENTHAL. 1975b. A review of the *Geomydoecus texanus* complex (Mallophaga: Trichodectidae) from *Geomys* and *Pappogeomys* (Rodentia: Geomyidae). Journal of Medical Entomology, 12:401-408.

PRICE, R.D., and R.A. HELLENTHAL. 1976. The *Geomydoecus* (Mallophaga: Trichodectidae) from the hispid pocket gopher (Rodentia: Geomyidae). Journal of Medical Entomology, 12: 695-700.

PRICE, R.D., and R.A. HELLENTHAL. 1979. A review of the *Geomydoecus tolucae* complex (Mallophaga: Trichodectidae) from *Thomomys* (Rodentia: Geomyidae), based on qualitative and quantitative characters. Journal of Medical Entomology, 16:265-274.

PRICE, R.D., and R.A. HELLENTHAL. 1980a. The *Geomydoecus oregonus* complex (Mallophaga: Trichodectidae) of the western United States pocket gophers (Rodentia: Geomyidae). Proceedings of the Entomological Society of Washington, 82:25-38.

PRICE, R.D., and R.A. HELLENTHAL. 1980b. The *Geomydoecus neocopei* complex (Mallophaga: Trichodectidae) of the *Thomomys umbrinus* pocket gophers (Rodentia: Geomyidae) of

Mexico. Journal of the Kansas Entomological Society, 53: 567-580.

PRICE, R.D., and R.A. HELLENTHAL. 1980c. A review of the *Geomydoecus minor* complex (Mallophaga: Trichodectidae) from *Thomomys* (Rodentia: Geomyidae). Journal of Medical Entomology, 17:298-313.

PRICE, R.D., and R.A. HELLENTHAL. 1981a. A review of the *Geomydoecus californicus* complex (Mallophaga: Trichodectidae) from *Thomomys* (Rodentia: Geomyidae). Journal of Medical Entomology, 18:1-23.

PRICE, R.D., and R.A. HELLENTHAL. 1981b. Taxonomy of the *Geomydoecus umbrini* complex (Mallophaga: Trichodectidae) from *Thomomys umbrinus* (Rodentia: Geomyidae) in Mexico. Annals of the Entomological Society of America, 74:37-47.

PRICE, R.D., and R.M. TIMM. 1979. Description of the male of *Geomydoecus scleritus* (Mallophaga: Trichodectidae) from the southeastern pocket gopher. Journal of the Georgia Entomological Society, 14:162-165.

ROTHSCHILD, M., and T. CLAY. 1957. Fleas, flukes and cuckoos: A study of bird parasites. New York: The MacMillan Company, 305 pp.

ROTHSCHILD, M., and R. FORD. 1964. Breeding of the rabbit flea (*Spilopsyllus cuniculi* (Dale)) controlled by the reproductive hormones of the host. Nature, 201:103-104.

ROTHSCHILD, M., and R. FORD. 1966. Hormones of the vertebrate host controlling ovarian regression and copulation of the rabbit flea. Nature, 211:261-266.

ROTHSCHILD, M., and R. FORD. 1969. Does a pheromone-like factor from the nestling rabbit stimulate impregnation and maturation in the rabbit flea? Nature, 221:1169-1170.

RUSSELL, R.J. 1968. Evolution and classification of the pocket gophers of the subfamily Geomyinae. University of Kansas Publications, Museum of Natural History, 16:473-579.

RUSSELL, R.J., and J.K. JONES, JR. 1956. The taxonomic status of *Geomys bursarius vinaceus* Swenk. Transactions of the Kansas Academy of Science, 58:512-513.

RUST, R.W. 1974. The population dynamics and host utilization of *Geomydoecus oregonus*, a parasite of *Thomomys bottae*. Oecologia (Berl.), 15:287-304.

SMITH, J.D. 1972. Systematics of the chiropteran family Mormoopidae. Miscellaneous Publication, Museum of Natural History, University of Kansas, 56:1-132.

SPICKA, E.J. 1981. Ectoparasites and other arthropod associates on two subspecies of plains pocket gophers: *Geomys bursarius illinoensis* and *Geomys bursarius missouriensis*. Canadian Journal of Zoology, 59:1903-1908.

TIMM, R.M. 1975. Distribution, natural history, and parasites of mammals of Cook County, Minnesota. Occasional

Papers, Bell Museum of Natural History, University of Minnesota, 14:1-56.

TIMM, R.M. 1979. The *Geomydoecus* (Mallophaga: Trichodectidae) parasitizing pocket gophers of the *Geomys* complex (Rodentia: Geomyidae). Ph.D. Dissertation, University of Minnesota, St. Paul. 124 pp.

TIMM, R.M., E.B. HART, and L.R. HEANEY. Karyotypic variation in pocket gophers (Geomyidae: *Geomys*) from a narrow contact zone in Nebraska. Mammalian Chromosomes Newsletter. (In press)

TIMM, R.M., and R.D. PRICE. 1980. The taxonomy of *Geomydoecus* (Mallophaga: Trichodectidae) from the *Geomys bursarius* complex (Rodentia: Geomyidae). Journal of Medical Entomology, 17:126-145.

VILLA, B., and E.R. HALL. 1947. Subspeciation in pocket gophers of Kansas. University of Kansas Publications, Museum of Natural History, 1:217-236.

WELTY, J.C. 1982. The life of birds. Philadelphia: W. B. Saunders Company. 754 pp.

WENZEL, R.L., V.J. TIPTON, and A. KIEWLICK. 1966. The streblid batflies of Panama (Diptera Calypterae: Streblidae); pp. 405-675. *In* Wenzel, R.L., and V.J. Tipton (eds.), Ectoparasites of Panama. Chicago, Field Museum of Natural History, 861 pp.

WERNECK, F.L. 1945. Os tricodectideos dos roedores (Mallophaga). Memorias do Instituto Oswaldo Cruz, 42:85-150.

WRIGHT, S. 1931. Evolution in Mendelian populations. Genetics, 16:97-159.

THE RISE AND FALL OF THE LATE MIOCENE UNGULATE FAUNA IN NORTH AMERICA

S. David Webb

Florida State University
University of Florida
Gainesville, Florida

The Clarendonian Chronofauna, noted for its rich ungulate fauna, emerged in North America during the Barstovian, reached its acme in the Clarendonian and declined by the end of the Hemphilliam, an overall span of some 10 million years. The rise and fall of this chronofauna were governed by the late Cenozoic trend toward cooler and drier climates at temperate latitudes. As forest biomes gave way to savanna, the richness of ungulate species and the ratio of grazers to browsers increased. Later, however, the spread of steppe conditions resulted first in the extinction of virtually all browsers, secondly in the decimation of grazers and finally in the wholesale destruction of the chronofauna.

The late Miocene ungulate fauna of North America shows remarkably detailed resemblances to the Recent ungulate fauna of Africa, despite their wholly independent origins. A useful basis for comparing them is a bivariate display of body size and molar volume for all species in a local fauna. Molar volume, calculated from the largest single lower molar, appears to be a more useful parameter than the hypsodonty index usually recommended. Preliminary data for African ungulates reveal a fairly regular relationship between body size and molar volume versus observed feeding modes (i.e., grazing, mixed feeding or browsing). The North American Miocene ungulate fauna resembles the African Recent fauna in species numbers (up to 18 in most local faunas), the array of presumed feeding types, and even the probable relative biomass values of the analogous species. The ecological vicar of Hippopotamus *was the extinct rhinocerotid,* Teleoceras, *shown to be a large amphibious grazer. Similarly the counterpart of the African genus* Giraffa *is the extinct North American camelid,* Aepycamelus, *a larger browser which could reach heights of five or six meters.*

The striking resemblances between these two independently evolved ungulate faunas suggest that they are structured by similar sets of selective constraints. An elaborate system of positive and negative feedback loops must effect the fitness of each ungulate species, and lead to a similar set of results in each continent. In the North American fossil record the increasing numbers of species, especially among grazers, and the progressive specialization of various species lineages show that the ultimate resemblance to the African ungulate fauna was the product of a long and complex evolutionary history.

INTRODUCTION

During the late Miocene, North America supported its richest array of large herbivores since the late Mesozoic. Quarries in continental sediments of this age regularly produce 15 or more species of herbivorous vertebrates with body weights exceeding five kilograms. This great assemblage of ungulates included a mixture of apparent browsers and apparent grazers; presumably it lived on a rich savanna composed of trees, grasses and herbs. The analogy between the North American savanna ungulate fauna of the late Miocene and the present African ungulate fauna has become a fundamental tenet of studies on this subject (Gregory, 1971; Webb, 1977).

The Miocene continental sediments in North America yield a reliable chronicle of the rise and fall of the large herbivore fauna. The consistent association of the same taxonomic groups during much of the Miocene warrants their recognition as a chronofauna. The Clarendonian chronofauna (Webb, 1969; Tedford, 1970) can be traced as a coherent association of species lineages from the Barstovian through the Clarendonian and well into the Hemphillian land-mammal ages in North

America, a span of about 10 million years. Thus, the Clarendonian chronofauna endured in North America nearly as long as the Permian vertebrate chronofauna discussed by Olson (this volume).

The purpose of this chapter is to suggest that the large herbivores of the Clarendonian chronofauna represented a coevolving set of primary consumer species which also had regular coevolving relationships with the producer species of the savanna flora. It will not be possible in this analysis to detail the interactions between the numerous species pairs involved in such a coevolving system. Instead this discussion focusses on the overall feeding structure of the savanna ungulate guild. It offers only the broad assumption that if a set of species persists for a long period of time, and each continues to evolve along a characteristic course, then the system must be stabilized by coevolutionary interactions. Positive and negative feedback loops must relate the fitness of one species to that of another. It is difficult to account for the persistence of such a chronofauna except by some scheme of coevolutionary interactions against a background of relatively stable environments.

Much of this chapter compares the North American savanna ungulate fauna of the late Miocene with the African savanna ungulate fauna of the present. The conclusion is that, though derived quite independently, at different times and from different faunal backgrounds, these two faunas became closely convergent evolutionary systems. If each fauna were stochastically assembled from available herbivorous stock, there should be only a weak correspondence between a Miocene New World ungulate fauna and a Recent Old World one. On the other hand, if each savanna ungulate system were stringently shaped by similar environmental constraints and each species by numerous coevolutionary interactions, then the resultant faunas should be closely comparable.

The obverse of coevolution is *coextinction*. Coextinction is the ultimate special case of evolutionary coadjustment involving the loss of two or more species linked by some form of mutual dependence. An important clue to a case of coextinction is the loss of the linked species at the same time and in the same region. During the decline of the Clarendonian chronofauna several possible examples of coextinction are recorded.

ORIGINS OF THE CLARENDONIAN CHRONOFAUNA

The Clarendonian chronofauna was based on "an aggregate of local faunas whose overall taxonomic similarity suggests that they formed part of a common zoogeographic region characterized by a particular suite of species which persisted for a geologically significant interval" (Tedford, 1970). In it, "the ungulates in particular show considerable stability in composition at the generic level... with most showing observable speciation and some geographic variation of populations, but no major change due to extinction or immigration."

The Clarendonian chronofauna arose during the Barstovian land-mammal age, about 15 million years ago. It extended through the Clarendonian and well into the Hemphillian some 5 million years ago. Its duration was thus at least ten million years.

The Clarendonian land-mammal age falls in the middle of the Clarendonian chronofauna (from about 12 to about 9 million years ago). Despite their shared proper names in this application, the "age" concept and the "chronofauna" concept have entirely different sets of values, definitions and purposes. The central significance of the age concept is to define spans of time on the basis of faunal change (*bio-stratigraphy*), whereas the latter addresses broad evolutionary and ecogeographic continuities within the succession of faunas. One concept is primarily geological, and the other is essentially biological. It is entirely appropriate that an age should have narrower and more rigorously defined chronological limits than a chronofauna.

The Clarendonian was improperly defined on the basis of a fauna, with only a vague querried reference to a rock-stratigraphic context (Wood et al., 1941). Webb (1969) partly rectified this problem by proposing as stratotype an old section published by Cummins and collected by Cope from about 10 miles north of Clarendon, Texas.

A more persistent problem of definitions concerns the lower limit of the Clarendonian Age, and this problem has some bearing on the origin of the Clarendonian chronofauna as well. The strata underlying those that produce the typical Clarendonian fauna in Texas are not fossiliferous and thus are of no practical value in establishing the lower limits of that age. In north-central Nebraska, however, the Minnechaduza fauna is closely comparable to the fauna from near Clarendon. It derives from the Cap Rock Member of the Ash Hollow Formation and occurs in the midst of a richly fossiliferous succession of late Miocene strata. This succession in north-central Nebraska therefore provides an excellent basis for defining the boundaries of the Clarendonian land-mammal age. Subjacent to the Ash Hollow Formation, the Valentine Formation consists of three members, in descending order the Burge, Devil's Gulch and Crookston Bridge members, each producing rich faunal sample (Skinner et al., 1968; Webb, 1969). Webb (1969) following Wood et al. (1941) included the fauna from the Burge Member of the Valentine Formation in the Clarendonian, but assigned the faunas from the lower members, especially the Niobrara River fauna from the Crookston Bridge member, to the Barstovian age. This assignment generally has been accepted, but remains rather arbitrary pending more detailed biostratigraphic studies on both sides of the proposed boundary. In any event the Clarendonian chronofauna spanned not only the Clarendonian age but also adjacent earlier and later ages.

ENVIRONMENTAL CONTROLS

The rise and fall of the Clarendonian chronofauna evidently were governed by secular changes in the environment (fig. 1). The trend during most of the Miocene and for that matter throughout the late Cenozoic was toward cooler and drier climates at temperate latitudes (Wolfe, 1978).

In the broad expanses of the midcontinent, regional climate had a direct influence upon the predominant vegetational formations; and the vegetation in turn provided the pervasive control on the evolution of terrestrial herbivores. In the course of the Miocene the prevailing midcontinental ecosystem changed from forest to park savanna to steppe. The large herbivore fauna responded in an evolutionary sense first by its rise to the acme represented by the Clarendonian chronofauna and later by its decline.

The lower part of fig. 1 records the number of ungulate genera known for each stage or substage during the Miocene in North America.

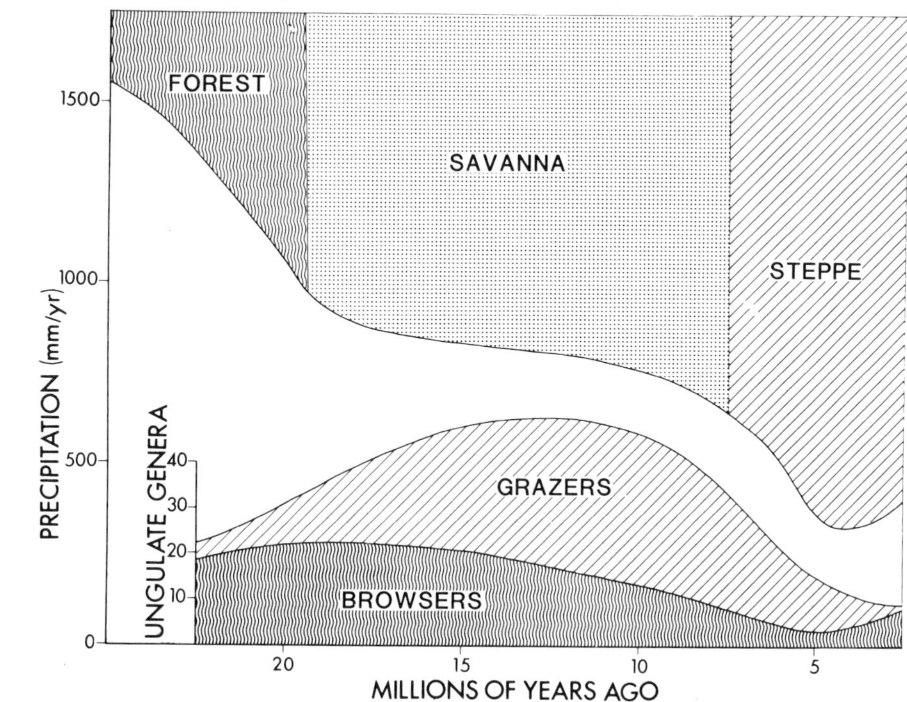

FIGURE 1. Increasing aridity during the Miocene produced a shift from predominantly forest to predominantly savanna and finally to predominantly steppe habitats. The fossil record of large ungulate genera in North America shows an increasing richness of grazing taxa. As steppe conditions came on, the number of browsing genera was severely reduced and even the number of grazing taxa was cut.

LATE MIOCENE SAVANNA FLORA

The savanna flora that presumably supported the Clarendonian chronofauna is not as well documented as the vertebrates that depended upon it. Nonetheless, since the great monograph of Elias (1942) the importance of grasslands in the midcontinent has been recognized. The published diversity of gramineae consists of only a few genera.

Voorhies and Thomasson (1979) for example cite only five genera in the late Cenozoic of Nebraska: *Berriochloa*, *Leersia*, *Nassella*, *Oryzopsis*, and *Panicum*. Presumably the true abundance of gramineaceous taxa has been concealed by preservational factors.

There has been an unfortunate tendency for "textbooks" to postulate a pure grassland in the late Miocene of North America. Chaney and Elias (1936) were well aware of the presence of moderately diverse forest settings, and regularly cited as an example their own study of the leaf flora from the Laverne Formation of Clarendonian age near Beaver, Oklahoma. Most importantly, MacGinitie's (1962) study of the rich Kilgore flora (of pollen and leaves) of probably late Barstovian age from the Valentine Formation in north-central Nebraska emphasized that even the interfluves did not support pure stands of grasses. Dense mesic riparian forests gave way to "open grassy forests ... pine-oak woodland on the interstream divides" (MacGinitie, 1962).

Presumably the situation at Kilgore is representative of the midcontinental flora during the late Barstovian and early Clarendonian. The continuity of the sediment blanket and the homogeneity of the terrestrial vertebrate fauna indicate broadly homogeneous conditions. In the late Clarendonian and early Hemphillian, however, conditions had surely begun to change considerably. The vertebrate faunal changes, to be discussed below, imply that the later floras of the High Plains will be found to be scrubbier than the Kilgore. One must be cautious, however, about interpreting the present limited evidence which consists mainly of fossil grass anthoecia from the Ogallala Formation (late Miocene) sediments. The siliceous reproductive structures of Gramineaceae, like the tough spherical endocarps of hackberries that often occur with them, had a much higher probability of being preserved in fluvial deposits than most other plant materials (Voorhies and Thomasson, 1979). In the absence of other detailed records we must accept the Kilgore flora as representative of the park savanna and riparian forest mosaic that blanketed the North American midcontinent in the late Miocene.

LATE MIOCENE SAVANNA UNGULATES

The large herbivores of the middle and late Miocene diversified in presumed evolutionary response to the expansion of savanna flora (fig. 1). During the Clarendonian age, North America supported 40 ungulate genera; by contrast there are now 11 ungulate genera in North America. In the ten million years from Hemingfordian through Barstovian time, the number of ungulate genera had doubled. At the outset, during Hemingfordian time, virtually all of the ungulates apparently were browsers, but they experienced no diversification whatsoever during the next ten million years, so that by the end of the Barstovian the proportion of browsers to all ungulates had fallen to less than half (fig. 1). In the late Clarendonian and early Hemphillian between 9 and 6 million years ago, the browsers were severely set back; and eventually the grazers were as well. Presumably the final decline to an almost modern low diversity of ungulates in late Hemphillian time (about 4.5 million years ago) represents the effects of a shift to steppe vegetation. It is largely a matter of definition whether one counts the mid-Hemphillian decline to 30 ungulate genera or the late Hemphillian drop to 18 genera as the termination of the Clarendonian chronofauna. The latter seems preferable if one considers the ungulates, because many grazing genera continue to the final Hemphillian extinction

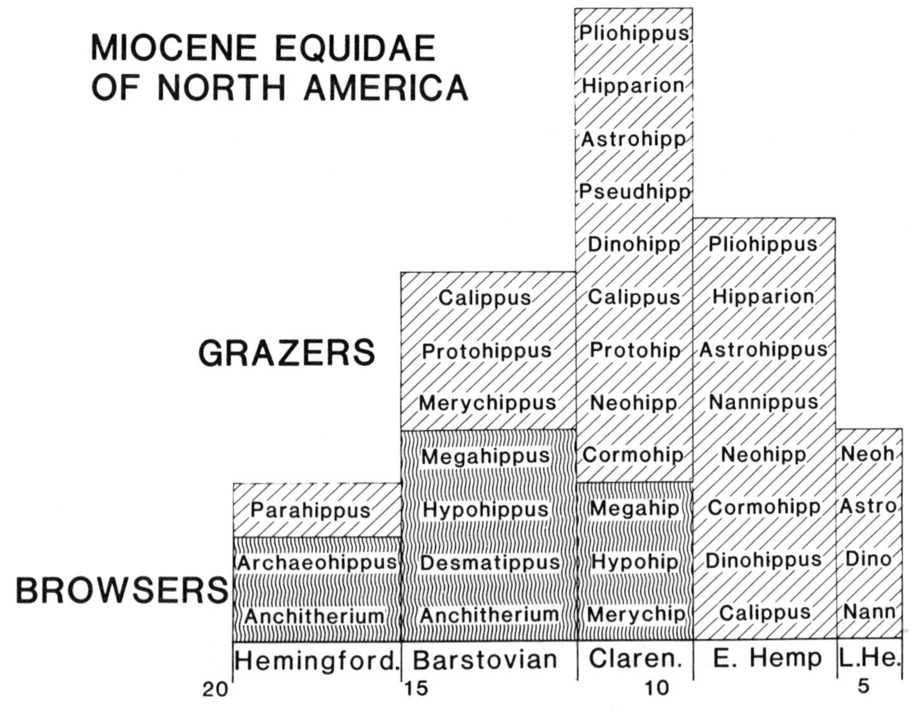

FIGURE 2. During the late Miocene, the number of horse genera in North America was quadrupled, principally as a result of diversification within native groups of grazers. (Their generic names are presented here). The browsers disappeared entirely about ten million years ago, and by five million years ago, only four genera of grazers were left. A rich savanna fauna (the Clarendonian chronofauna) prospered during the Barstovian, Clarendonian and early Hemphillian, between about 15 and about 6 million years ago.

episode. It is noteworthy that by the late Hemphillian, the number of ungulate genera had fallen even below its "beginning" level of 22 genera in the late Hemingfordian, and furthermore, that virtually all were grazers by then.

The well-known record of fossil horses, most of which evolved in North America, plays a central role in the history of the Clarendonian chronofauna. Fig. 2 summarizes the

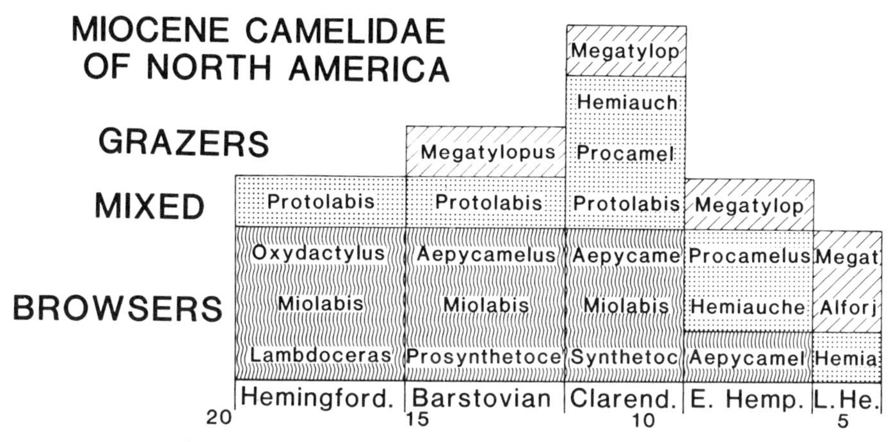

FIGURE 3. The number of native camelid genera in North America more than doubled during the late Miocene. The net change was from predominantly browsing forms to predominantly grazers. Three successive genera of Protoceratidae are also included in the lower left part of this figure.

history of equid genera during the middle and late Miocene. The number of genera doubles each five million years, from late Hemingfordian into early Clarendonian time (i.e., from 20 to 10 million years ago) and most of the diversification during that interval was among the grazing taxa. A similar pattern is found in the camelidae and a related family, the protoceratidae (fig. 3). In these families, however, the rate of diversification was not so high, and some of the diversification may have been to mixed-feeders rather than to strict grazers. These two figures may be taken as representative of the whole array of ungulates of the Clarendonian chronofauna.

INDEPENDENCE OF THE NORTH AMERICAN AND AFRICAN UNGULATE FAUNAS

The fossil record of Miocene land mammals from North American and adjoining continents is sufficiently well-known to assert that the geographic distributions of most genera can be tracked through time. In the present context it is important to know that the diversification of the Clarendonian chronofauna was essentially a North American achievement. The major ungulate families just noted, namely the Equidae, Camelidae and Protoceratidae, had long histories confined to, or for the Equidae at least centered in, North America. For the other two families of Perissodactyla (Tapiridae and Rhinocerotidae) and the other six families of Artiodactyla (Tayassuidae, Merycoidodontidae, Gelocidae, Dromomerycidae, Moschidae, and Antilocapridae) all but one of the genera or even subfamilies that contribute to the Clarendonian chronofauna were already established exclusively in North America at least by middle Miocene time. The one exception is the rare gelocid genus *Pseudoceras* which appeared in Clarendonian time. Indeed, of the Clarendonian artiodactyl families, only the Gelocidae and Moschidae are not exclusively North American. And finally two other ungulate families, the Mammutidae and Gomphotheriidae, came near the beginning of the Barstovian and were well established as North American genera as the Clarendonian chronofauna began to diversify. *Platybelodon*, a shovel-tusked genus of Asiatic ancestry is another exception since it appears in North America in some Clarendonian and Hemphillian sites. Nonetheless, nearly all of the 40 ungulate genera (in 13 families) that make up much of the Clarendonian chronofauna were native stocks that evidently evolved in place (Webb, 1977).

Similarly, it is important for this discussion to establish that the Recent African ungulate fauna had its own independent roots in the Old World. A brief analysis at the family level will suffice. The extraordinary richness of the Artiodactyla, especially the Bovidae, accounting for more than 80% of African ungulate species, includes no families in common with the American Miocene ungulates: there are Suidae instead of Tayassuidae, Giraffidae instead of Dromomerycidae, Tragulidae instead of Gelocidae, Bovidae instead of Antilocapridae and Camelidae; and Hippopotamidae quite without a North American artiodactyl counterpart. The distinctions are no less valid in the Proboscidea; only the Elephantidae are now present in Africa, whereas the older families (including a distant collateral branch of elephant ancestry among the Gomphotheriidae), occurred in the Clarendonian chronofauna. The African and North American savanna ungulate faunas share two families among the perissodactyls. Nevertheless, the genera of Rhinocerotidae in question had long independent histories in each hemisphere, and are usually placed in distinct tribes or subfamilies. And among Equidae, there is a directly traceable heritage from African zebras and asses to some late Tertiary North American genus, probably *Dinohippus*. The other ungulate orders in Africa such as Hyracoidea and Tubulidentata have independent histories from any possible New World relatives since at least the Eocene.

TO PARTITION AN UNGULATE FAUNA

Assuming that ungulates are essentially herbivorous, the fundamental division among feeding types traditionally has been between "browsers" and "grazers". During the last twenty years many outstanding studies of African ungulate feeding ecology have refined these concepts (Lamprey, 1963; Bell, 1971; Jarman and Sinclair, 1979; and many others) and now generally recognize three broad feeding categories: *roughage-feeders* (approximately the same as "grazers" but not implying strictly grasses); *juicy-herbage-feeders* (including trees, shrubs and fruit, approximately the same as "browsers"); and *intermediate-*

feeders (often including a mixture of grasses and dicotyledons, often seasonally changing emphasis). While continuing to use the familiar terms "grazers" and "browsers" we here also insert the third term "mixed-feeders".

It has long been supposed that many ungulates with the feeding preferences of "grazers" have *hypsodont* (high-crowned) dentitions and that "browsers" have *brachydont* (low-crowned) dentitions. Hypsodonty is ordinarily defined by an index of crown height to occlusal area (or the average of the two dimensions) for the tallest grinding tooth of over 1.0 (Van Valen, 1960). On the other hand brachydont teeth are taken as those with an index of less than one. Although this view has the appeal of simplicity and the authority of frequent repetition, it does not seem to bear close scrutiny. Because of the empirical ecological studies in modern African ungulates, it is a simple matter to test the efficacy of hypsodonty indices. Tieszen and Imbamba (1980) directly confirmed the dietary preferences of ungulate species by C^{13} "labelled" fecal samples. The African ungulates evidently have not read the treatises on hypsodonty, for several species diverge from the predicted results. For example, *Hippopotamus amphibius* make a practice of grazing on tall grasses near their aquatic places of residence, yet the unworn crown height of each molar is less than its anteroposterior length, thus relegating this species to the brachydont category on the basis of its hypsodonty index.

The patently incorrect result in the case of *Hippopotamus amphibius* suggests a better approach. Clearly the long molars of this species provide the complex enamel folds needed to triturate its coarse fodder at least as effectively as do the tall selenodont teeth of a buffalo. Presumably crown height must be considered in terms of a large animal's longevity as well as in terms of its masticating ability at any one instant. Such considerations suggest that *the most important dental measurement affecting the overall fitness of a grazing animal is the volume of its grinding teeth.*

In the following analyses I have measured the volume of only *the largest single lower molar* for each ungulate species (table 1). Clearly this procedure ignores the importance of the whole cheek tooth battery. In practice, however, the whole-battery approach would require many extra calculations and also a large diverse-aged sample of dentitions so that each tooth could be measured at an unworn stage. Furthermore the results would not be coherent. While many ungulate groups with molariform premolars, such as Elephantidae and Recent Equidae, are clearly revealed as grazers on the basis of the volume of the single largest lower molar, others such as the extinct anchitheriine horses and the African members of the Rhinocerotidae are browsers. If the volume of the whole cheek-tooth battery of a browsing rhino were compared with the whole cheek-tooth battery of a similar-sized grazing ruminant, (in which group molariform premolars never occur), the results would not accord with feeding observations. Nonetheless, within the ruminants, single molar values give concordant results. Presumably the difficulty lies in the very important differences between ungulates with digestive specialization of the caecum and those with digestive specialization of the rumen (Janis, 1976). Thus it appears that single molar volume distinguishes between grazers and browsers among ungulates with specialized caecal digestion and molariform premolars on about the same basis as it does between grazers and browsers among ungulates with specialized rumen digestion and non-molariform premolars.

A second factor of primary importance to the feeding

Table 1. Molar Volume and Feeding Modes in Recent African Ungulates (in mm³).

	Width	*Length*	*Height*	*Volume*
Grazers:				
Equus burchelli	12.6	27.1	82.8	28,273
Hippopotamus amphibius	29.2	41.2	37.7	45,355
Syncerus caffer	16.1	27.5	24.8	10,980
Kobus kob	9.3	11.6	22.8	2,381
Hippotragus equinus	14.0	28.1	16.3	6,412
Oryx oryx	10.9	23.8	20.4	5,292
Damaliscus dorcas	8.6	16.5	5.4	2,185
Damaliscus korrigum	10.5	18.0	23.5	4,442
Connochaetes taurinus	11.2	23.9	19.0	5,086
Gazella thomsoni	6.0	11.9	11.0	785
Mixed-Feeders:				
Tragelaphus scriptus	6.4	11.2	13.3	953
Taurotragus oryx	16.2	24.7	22	8,803
Redunca sp.	7.4	10.0	20	1,480
Aepyceros melampus	8.9	15.5	16.2	2,235
Gazella granti	8.4	17.3	12.6	1,831
Browsers:				
Tragelaphus strepsiceros	12.4	22.8	24.1	6,814
Diceros bicornis	32.6	54.3	48.7	8,621
Giraffa camelopardalis	22.2	28.8	16.5	6,560

strategy of an ungulate is its body mass. Clearly a Thomson's Gazelle weighing less than 10 kgs need not process as much food as an elephant weighing 300 times as much. On the other hand, the elephant does not have to process 300 times as much food as the Thomson's Gazelle, because it has a lower relative metabolic rate. (Generally in mammals energy expenditures are proportional to the 3/4 power of body weight). One would expect tooth volume to bear the same exponential relationship to body weight, if tooth volume is directly related to the volume of fodder it can triturate, and volume of fodder is directly related to metabolic work. Longevity might influence this relationship also, but to judge from Western's (1979) discussion, this may not be critical in ungulates. In any event, since the relationship between body size and masticatory efficiency is of first order importance, one must consider body weight as well as tooth volume in any analysis of mammalian herbivore feeding strategies.

Table 1 presents a preliminary sample of molar volume data from various East African ungulate species and fig. 4 displays the relationship between molar volume, body weight and known feeding strategy for various East African ungulates. The regression of molar volume on body weight falls at about 0.75 as expected from the energetics considerations discussed above.

COMPARISONS BETWEEN TWO SAVANNA UNGULATE FAUNAS

If one now returns to the ungulates of the Clarendonian chronofauna it is possible to partition them in the same manner as the Recent African ungulate fauna. The molar volume and estimated body weight data for the array of North American Miocene ungulate species are illustrated in fig. 5.

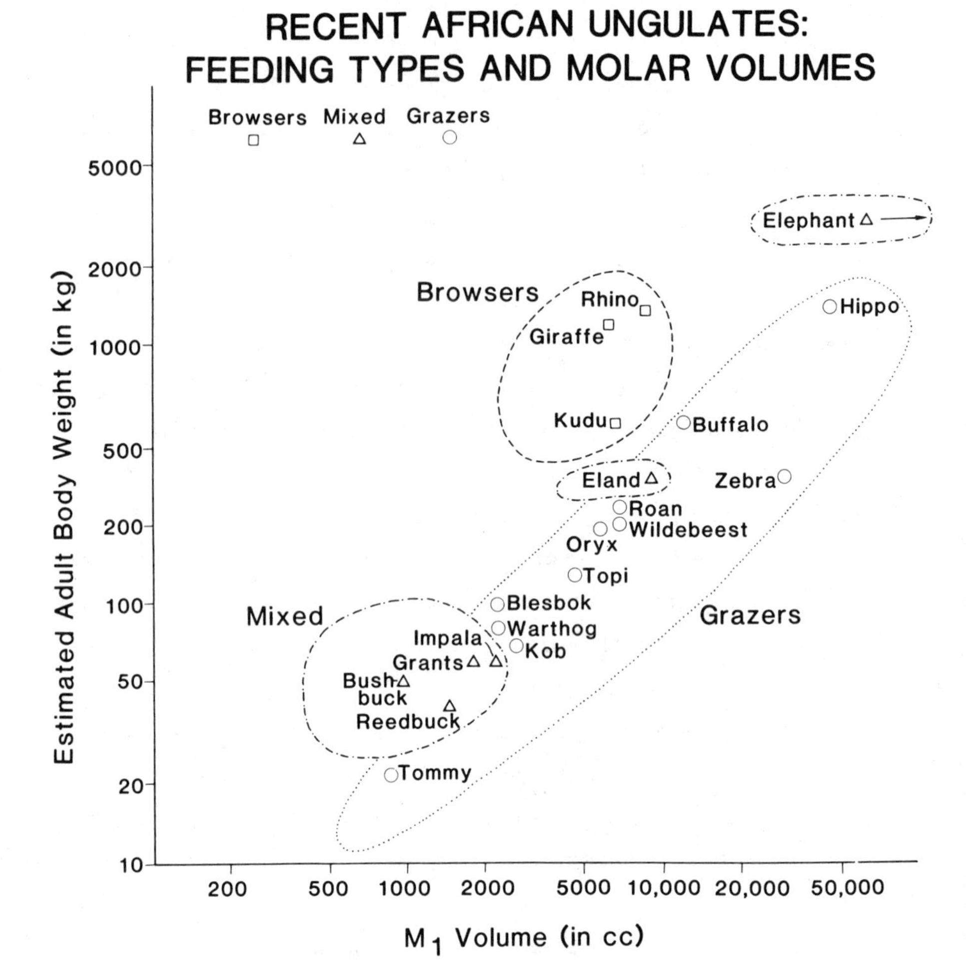

FIGURE 4. Selected living ungulate species from African savannas arrayed by body weight (ordinate) and volume of largest molar (abscissa). A general correlation between relative tooth volume and feeding type exists, with grazers making up the large array of relatively large toothed species.

It should be noted here that these data are preliminary and require more extensive sets of measurements. As for the Recent African data they are based only on specimens available at the Florida State Museum and represent only a pilot study.

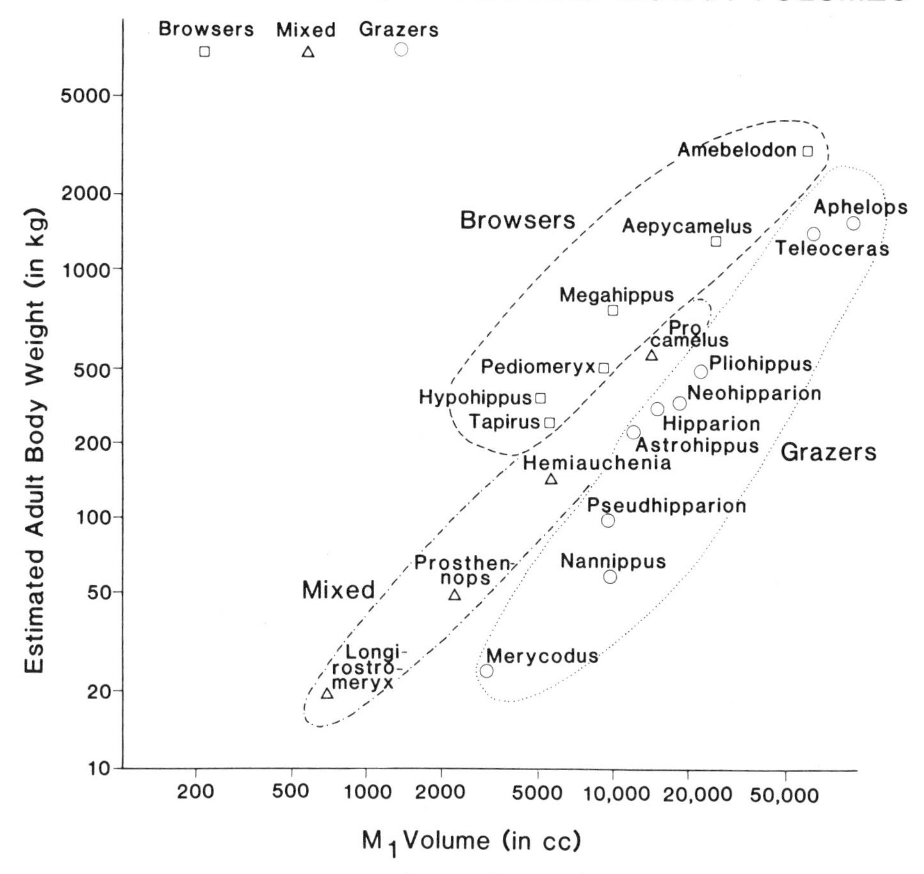

FIGURE 5. An array of North American Miocene ungulates on the same axes as in fig. 4, interpreted as having similar feeding strategies as in fig. 4.

The fossil species' body weights are estimated by comparing various linear measurements with living equivalents and then taking their approximate body weights. Various more objective systems could be employed (see Radinsky, 1982). Possibly the most useful relationship for estimating fossil ungulate body weights is that of MacMahon (1975) which shows a remarkably tight regression of metatarsal cross-section on body weight for a broad array of living ungulates. Such refinements will be necessary as this kind of analysis is extended to more complete data sets.

The primary result of this preliminary survey is that the array of the Clarendonian ungulate fauna is broadly comparable to that of the Recent African ungulate fauna. Similar divisions into "grazers", "mixed-feeders" and "browsers" may be made on the basis of molar volume and estimated body weight. *Evidently these separate evolutionary experiments with savanna ungulate faunas yielded broadly similar results with respect to relative numbers of the three major feeding types.*

INDEPENDENT EVIDENCE OF EXTINCT UNGULATE ROLES

Once the fossil ungulate fauna is partitioned as in fig. 5, it may be possible to test the feeding strategies of various species from independent fossil evidence. Such tests may consist of structural correlatives not directly related to body size or molar volume; or they may be derived from environmental or taphonomic circumstances. Two examples may serve as partial tests of the foregoing scheme of feeding strategies.

The largest presumed browser in the Clarendonian chronofauna is the genus *Aepycamelus*. It is clearly derived from autochthonous camelid stock in the Miocene of North America. It is known from most local faunas of Barstovian, Clarendonian and early Hemphillian age by a string of nominal species including *A. alexandrae*, *A. giraffinus*, *A. bradyi* and *A. major*. The principal trends in this approximate lineage are increasing size and exponentially increasing lengths of neck and limbs. The late Clarendonian species *A. major* provides the data for fig. 5. Clearly the adaptive role of *Aepycamelus* suggested by these data is that of a large browser, the ecological vicar of *Giraffa camelopardalis* in Africa. The most obvious giraffe-like features are long limb proportions and -- presumably the ultimate adaptive point -- great head height. In a partial skeleton of *Aepycamelus giraffinus* Matthew (1901) each of the cervical vertebrae and each of the hind limb elements exceeded that of a large northern giraffe.

FIGURE 6. Restored view of Aepycamelus giraffinus, *late Miocene giraffe-camel of North American savannas, convergent with African giraffe.*

The estimated shoulder height is 3.5 meters (fig. 6).

The largest presumed grazers in the Clarendonian chronofauna (as partitioned in fig. 5) are members of the genus *Teleoceras*. The genus has its roots in the Hemingfordian and is a North American endemic member of the family Rhinocerotidae. The data for molar volume and estimated body weight were taken from a mounted skeleton and a large sample of *Teleoceras fossiger* from the Love Bone Bed (late Clarendonian) in Florida (Webb, MacFadden and Baskin, 1981). Clearly the ecological vicar of *Teleoceras* in the East African savanna fauna is the semiaquatic grazer, *Hippopotamus amphibius*.

Three independent lines of evidence support the interpretation that *Teleoceras* played the adaptive role of a hippo in the Miocene of North America. First, the body proportions and body size are remarkably close between the brachypotheriine rhinocerotid from North America and the hippopotamus from Africa (fig. 7). Shoulder height in a large *Teleoceras fossiger* is about 1.5 meters as in a large hippo, and the limb proportions (though not the cranial proportions) also agree well with those of a hippo.

The other two lines of evidence bearing on the ecology of *Teleoceras* are most clearly developed from the work of

FIGURE 7. Restored view of Teleoceras fossiger, *late Miocene rhinocerotid of North American savannas, convergent with African hippo.*

Mike Voorhies and University of Nebraska field parties at the fabulous Poison Ivy Quarry in northern Nebraska. Originally a lake, the quarry was filled in rapidly by airborne ash which preserved the fauna completely and without transportation. The most conspicuous sample was a herd of *Teleoceras major* including cows with calves at their sides, animals which evidently had dwelt in the lake. Furthermore, careful excavation, especially along the hyoid bones of the throat, revealed numerous anthoecia of the grass genus *Berriochloa* associated with the digestive tract of *Teleoceras* specimens (Voorhies and Thomasson, 1979). Thus, *Teleoceras*, like *Hippopotamus*, evidently lived in the water but grazed on adjacent dense grasslands.

The population structure of a fossil species sample may also shed light on its adaptions. For example, analysis of the large *Teleoceras* collection from the Love Bone Bed in Florida shows a much lower juvenile mortality rate than the sample of *Aphelops*, another rhinocerotid from the same site. In this and other aspects of its population structure the *Teleoceras* sample resembles the living hippopotamus, whereas the *Aphelops* population structure closely compares with that of the living black rhinoceros (Wright, pers. comm.).

In another population study from the Love Bone Bed Hulbert (1982) developed evidence that the presumed grazing horse *Neohipparion* probably made mass migrations as do zebras and other savanna grazers today. At this stream site, the newborn foal class is substantially underrepresented suggesting that they were born elsewhere, presumably in upland prairies during the spring rainy season. Such seasonal migration also explains the discreteness of the subsequent dozen age classes in the perennial stream sample from the Love Bone Bed.

A major task of paleobiologists is to develop hypotheses and practical tests of ancient adaptations such as these. The fact that several tests from independent lines of fossil evidence have confirmed the feeding adaptations hypothesized for *Aepycamelus* and *Teleoceras* does not of course exonerate the scheme for all taxa as presented above. On the other hand, these preliminary results encourage further investigations with these working hypotheses.

BIOMASS AND ABUNDANCE IN TWO UNGULATE FAUNAS

One of the great achievements of African large mammal ecologists has been a coherent set of biomass estimates for a number of major reserves. Furthermore reasonable hypotheses have been generated to explain variance between ungulate biomass estimates from various surveyed areas (Western, 1980). Fig. 8 presents the biomass (as a percentage of the total standing crop) for each ungulate species in the Serengeti-Mara Ecosystem (Sinclair and Norton-Griffiths, 1979), in the same framework of feeding strategies as fig. 4. Table 2 provides similar data sets from several of the best-known reserves in Africa. In general, one notes that the biomass of browsers totals less than 10% of all ungulate biomass; that mixed-feeders account for about 15 to 30%; and that grazers predominate with some 60 to 80% of the total.

The rich fossil record of the Clarendonian chronofauna also provides potential evidence of species abundance and therefore of relative biomass. It would seem a simple matter, at least in principle, to collect a sample of bones and teeth from a single major quarry and identify them all. Such data would give the relative abundance of elements by species at the site of deposition, but several potential biases of

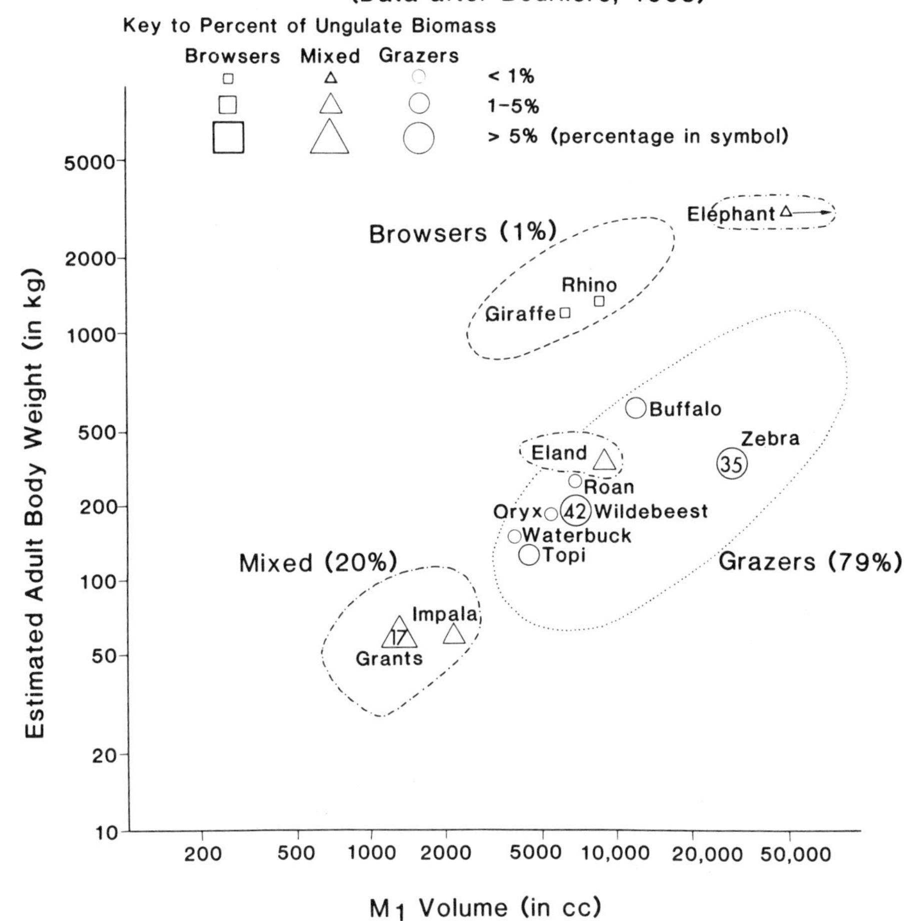

FIGURE 8. Serengeti ungulate biomass, percent of total distributed by species, presented in the same array as fig. 4. Note that browsers make up only 1% of total ungulate biomass, mixed-feeders about 20% and grazers nearly 80%.

uncertain magnitude separate them from living biomass data. Such problems have been the subject of extensive discussion and experiment (e.g., Shotwell, 1958; Voorhies, 1969; Behrensmeyer and Hill, 1980).

In floodplains systems, body size may be the most important factor influencing the probability of deposition of a fossil element. In an important control study of modern ungulate bone deposition in the Amboseli Park, Behrensmeyer and Boaz (1980) found that "Carcasses of individuals of greater than 100 kg average adult weight are recorded in greater than expected frequencies..., while carcasses of individuals between 15 and 100 kg average adult weight are recorded in smaller than expected frequencies." This study showed a regular exponential relationship between body size and bone frequency and that "size-related taphonomic processes in the Amboseli system include destruction by carnivores, trampling and weathering."

Although several other major sampling problems exist, it seems likely that in a fluviatile system, the most important biases are negatively correlated with size. A crude sampling correction may be obtained by dividing the fossil element frequency for each species by its estimated body weight. On the other hand, to estimate percentage of standing crop biomass, each frequency must be multiplied by estimated body weight. These two operations thus cancel one another and lead to a remarkably simple result: *Among ungulate species in a fluviatile system the relative frequency of preserved elements yields a crude estimate of relative biomass.*

Table 2 and figs. 9 through 11 present the relative frequencies of ungulate species in five major quarries representing the Clarendonian chronofauna. These samples include the time span from 6 to 12 million years ago. In each of these, the relative biomass estimate for browsers is generally less than 10% (except that it approaches 20% at the Love Bone Bed in Florida); about 10 to 30% for mixed-

Table 2. Estimated Biomass Distribution in Ungulate Samples

I. AFRICA: SITE	RAINFALL (mm/yr)	No. of Spp.	Browsers %	Mixed %	Grazers %
Rwindi	863	10	--	21.7	78.3
Nairobi	844	18	6.6	24.5	69.0
Amboseli	350	9	8.1	32.9	59.0
Serengeti	803	15	1.2	19.3	79.5

II. NORTH AMERICA: SITE	AGE (mybp)	No. of Spp.	Browsers %	Mixed %	Grazers %
Burge	12	16	5.2	18.0	76.8
Minnechaduza	10	16	2.8	10.4	86.8
Love Bone Bed	9	18	19.2	26.1	54.7
Boardman	7	8	4.4	26.2	69.4
Coffee Ranch	6	13	7.3	12.5	80.2

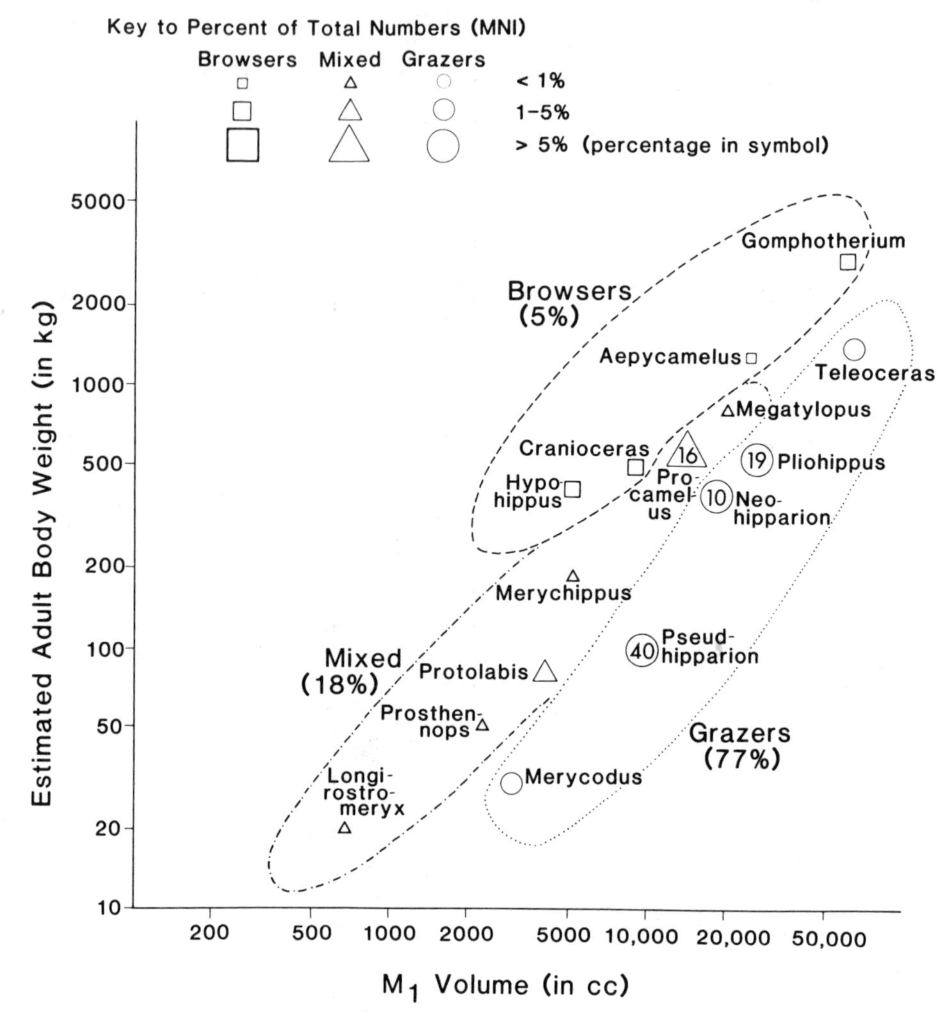

FIGURE 9. Burge Quarry ungulate percentages, possibly approximating biomass percentages (see text). Note that browsers account for only about 5% of total ungulate count, mixed-feeders about 20% and grazers about 80%.

feeders; and about 55 to 85% for grazers. The overall resemblance to the weighting of these same feeding categories for the African ungulate fauna is remarkable.

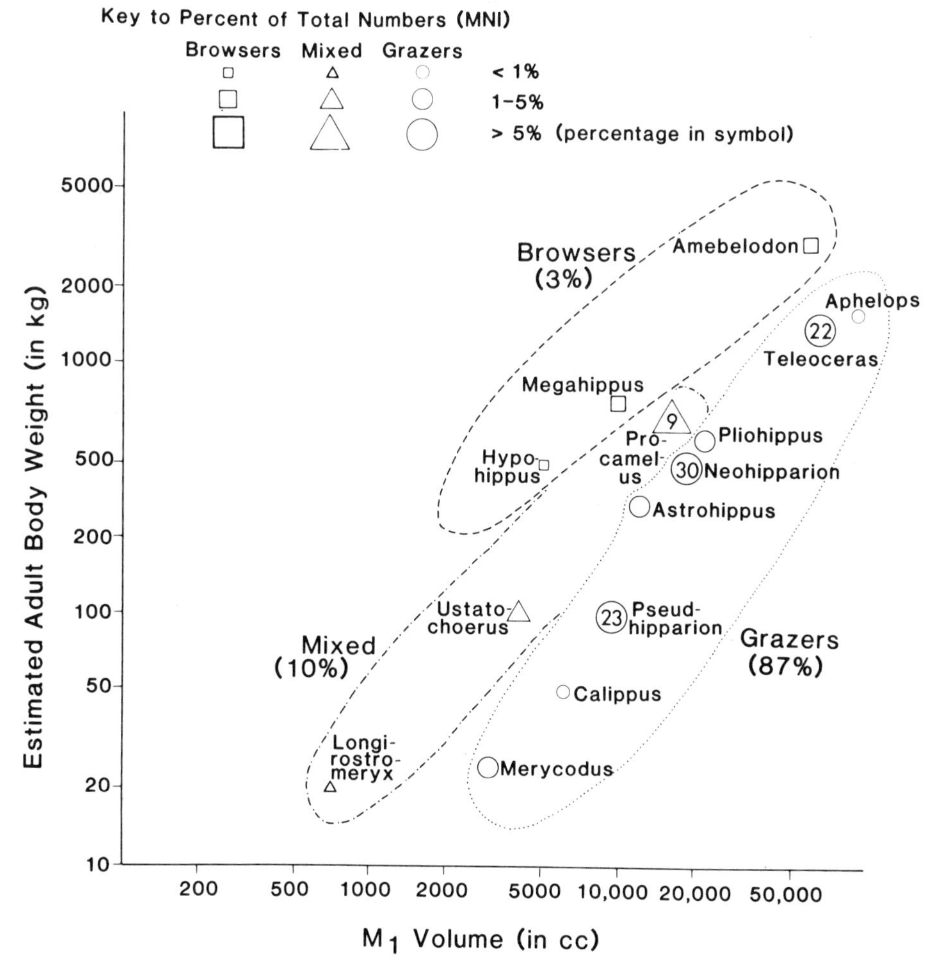

FIGURE 10. Little Beaver B Quarry ungulate percentages. Note percentages assigned to different feeding strategies. MNI = minimum number of individuals.

A final point of comparison is the total number of ungulate species living in a given region. The functional units here are game reserves in the East African savannas, and major quarry samples in the Miocene fluviatile sediments of North America. In each case several adjacent habitats

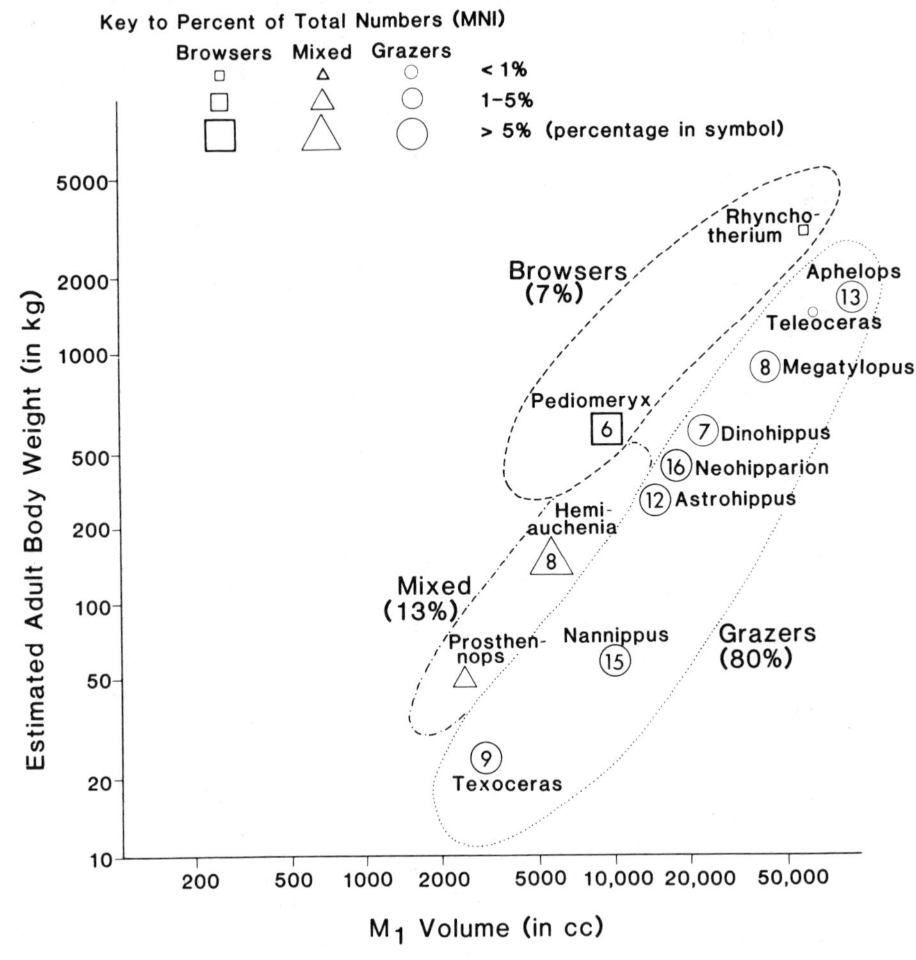

FIGURE 11. Coffee Ranch Quarry ungulate percentages; note percentages assigned to different feeding strategies. MNI = minimum number of individuals.

are probably sampled, yielding beta diversity samples in the sense of Whittaker (e.g., Cody, 1975). The further point of resemblance is striking, for as noted in table 2 the range in both ungulate faunas is from 8 or 9 to 18 species. Thus in the array of feeding strategies, in the biomass devoted to each of these strategies and in the total number of broadly

Table 3. Extinction Patterns in the Clarendonian Chronofaunal Demise.

Late Clarendonian	*Early Hemphillian*	*Late Hemphillian*
PROBOSCIDEA		
		Gomphotherium
		Amebelodon
		Platybelodon
PERISSODACTYLA		
Rhinocerotidae		
Peraceras		Aphelops
		Teleoceras
Equidae		
Hypohippus	Cormohipparion	Pseudhipparion
Megahippus	Hipparion	Calippus
Merychippus	Pliohippus	Neohipparion
Protohippus		Astrohippus
ARTIODACTYLA		
Tayassuidae		
		Macrogenis
Merycoidodontidae		
Ustatochoerus		
Protoceratidae		
	Synthetoceras	Kyptoceras
Camelidae		
Miolabis	Aepycamelus	Alforjas
Nothotylopus	Procamelus	
Protolabis		
Gelocidae		
	Pseudoceras	
Moschidae		
Blastomeryx		
Longirostromeryx		
Dromomerycidae		
		Pediomeryx
Antilocapridae		
Merycodus	Sphenophalos	Texoceros
Proantilocapra	Ottoceros	Ilingoceros
		Herameryx
Bovidae		
		Neotragoceras
13 genera	*9 genera*	*17 genera*

sympatric species supported, the Recent African and the Miocene North American ungulate faunas display remarkable convergent resemblances.

COEXTINCTION IN SAVANNA UNGULATES

The obverse of coevolution is coextinction. If the members of two species are regularly symbiotic, then a decline or range restriction in one species will surely have detrimental effects on the other species. For example, Bell (1971) and Sinclair (1979) have shown that during the dry season migrations of ungulates in the Serengeti-Mara Ecosystem, mass-grazing zebra and wildebeest herds "clear the way" and stimulate new growth in the *Themeda-Pennisetum* grassland which is subsequently utilized by the selective-grazing little Thomson's Gazelles. Decimation of populations of zebra, of wildebeest or of both in that region would surely reduce the fitness of many members of the Thomson's Gazelle population.

Although it may rarely be possible to shed direct light on such interactions among fossil species, scrutinizing cases of synchronous extinctions may provide the first hint of a true coextinction. Such circumstantial evidence may be strengthened by other ecological or geographic coincidences between two taxa that later become coextinct.

Table 3 lists three sets of synchronous ungulate extinctions that brought about the decline of the Clarendonian chronofauna. Clearly it is unlikely that all of these approximately synchronous last records could be closely linked symbiotically. Indeed, most of the late Clarendonian extinctions represent browsing taxa, for which the most likely explanation is a restriction in forest vegetation, *Aepycamelus* and *Pediomeryx*, the two largest browsers being the principal survivors. More detailed evidence of coevolutionary interaction is needed.

Several possible coextinctions among grazing taxa in the early Hemphillian are most suggestive. For example, one may postulate interspecific facilitation between the large grazing herds of horses such as *Cormohipparion* and the abundant medium-sized antilocaprids such as *Sphenophalos*. Another pair of extinctions are the small but abundant grazing horses, *Calippus* and *Pseudhipparion*. Both genera appear last in the High Plains in the late Clarendonian, but persist in the Gulf Coastal Plain through the Hemphillian. In that region, however, both become extinct synchronously in the very late Hemphillian. Possibly further circumstantial evidence will link *Calippus* and *Pseudhipparion* as a case of coextinction.

In broader terms, a series of climatic deteriorations effectively destroyed the richness and apparent stability of the late Miocene savanna biota. Whatever network of coadaptations had previously engendered and maintained the savanna flora and its rich ungulate fauna collapsed with relative rapidity before the Tertiary Period had ended.

CONCLUSIONS

Extensive late Miocene savannas in North America supported a richly varied ungulate fauna. Following its origin from a limited number of browsing stocks, it persisted with few major changes for about 10 million years. A typical quarry sample from this Clarendonian chronofauna may include up to 18 species. Estimates of the biomass indicate about 80 percent grazers, 15 percent mixed feeders, and about 5 percent browsers. The fauna was decimated and then largely destroyed when the savannas gave way to steppe.

A central purpose of this essay has been to test the proposition, implicitly accepted by many paleontologists, that the Clarendonian chronofauna produced "an array of taxa fully comparable to that in African savannas today" (Webb, 1977). Several different lines of evidence seem to warrant this conclusion. First, each of these great ungulate faunas yields between 10 and 18 species in a given sampling area. The array of grazers, mixed-feeders, and browsers, as recognized by the molar volume at a given body size, is closely comparable also. Approximations of biomass distribution within this array showed less than 10% of the standing crop consisting of browsers; 15 to 30% made of mixed-feeders; and about 60 to 80% of the total ungulate biomass composed of grazers. At least some of the grazers relied on regular seasonal migrations. The largest grazer in each fauna was a short-legged amphibious grazer with very low juvenile mortality: in North America this role was played by *Teleoceras*, an endemic rhinocerotid, whereas in Africa the ecological vicar is *Hippopotamus* from a different family and order. The tall browser in North America is *Aepycamelus*, a native camelid, whereas in Africa the genus *Giraffa* is derived from a distinct suborder on that continent. Thus two ungulate faunas, evolving in different hemispheres during different geologic epochs, attained remarkably convergent features.

The force of this comparison is to raise the question as to what processes produced such similar ungulate faunas. Clearly they are not randomly assembled. As outlined above, each fauna was derived from an altogether different background. And each fauna was considerably modified from its origin to its "mature" state. Presumably the selective forces that shaped each of these faunas produced these convergent results.

Before proceeding further, it may be well to note that the resemblances between these two rich ungulate faunas do not amount to complete identity. Janis (pers. comm.) has warned against the "Serengeti fallacy" which would equate all aspects of these two great ungulate faunas. It would indeed be astonishing if two faunas with no species in common were identical in every other way. The problem, then, is to factor out which features of these convergent faunas have been selected in a convergent manner, and which remain "true" to their heritage. The principal resemblances are in the numbers of species and their distribution by feeding types, body sizes, and biomass.

The fundamental environmental feature shared by the North American and African ungulate faunas is a vast savanna setting. The extraordinary richness of large herbivores in each continent may be traced to the optimal mixture of grasses, herbs, and trees. From the North American fossil record it is clear that more forested situations supported a more limited fauna of browsing ungulates, whereas more open situations supported a more limited fauna of grazing ungulates. In the optimum savanna setting, feeding modes were presumably the critical adaptations that were subjected to the most severe competitive constraints. In East Africa Sinclair (1975) showed that a protein shortage during the dry months was probably the limiting factor for most herbivores. There, some large ungulates with specialized caecal digestive systems, notably zebras, survive by sheer nutritional inertia (Janis, 1976). Some small grazers, such as Thomson's gazelles, become highly selective, consuming only the richest remaining bits of foliage. Others, such as wildebeest, survive by migrating to wetter regions. Intense pressure must drive each species to find its most efficient means of obtaining essential vegetational resources during the dry season.

The present study sheds little direct light on the detailed nature of such competitive interactions between savanna ungulates. Rather it suggests only the broad patterns of diversity and feeding modes that have resulted from two essentially replicate experiments. In Africa and in North America about 10 to about 18 species of ungulates are found in a given savanna area (or quarry sample). If each species interacts with every other, as might be expected among large mobile herbivores, then the number of interacting species pairs would equal $\frac{n(n-1)}{2}$, or between 45 and 153 such interactions. Presumably strong selection reduced the intensity of competition through rapid divergence in feeding modes early in the evolution of each fauna. There may also have been strong selection favoring sets of species that interacted positively, as when members of one species facilitate the feeding of another, or when predator alarms by members of one species alert another. Such felicitous interactions are well documented in Africa where zebras and wildebeest facilitate the selective feeding of Thomson's gazelles during dry season migrations (Sinclair and Norton-Griffiths, 1979). Thus strong selection must have favored finely tuned methods of partitioning vegetational resources among ungulate species. The foregoing comparison of two separate savanna systems suggests that the results of such coadaptive processes are remarkably predictable.

It is worth remembering that these events took place not in ecological but in evolutionary time. The Clarendonian chronofauna developed slowly, along with the North American savanna, through a period of some ten million years with very limited influence from any allochthonous taxa. A similar history may be inferred from the available record of the

African fauna. As Sinclair (1979) observes: "The Serengeti ecosystem appears on this evidence to have been very similar to present conditions for a long time, at least a million years and probably much longer. The changes in climate that have occurred were not great and took place gradually. Consequently, the interrelationships between organisms that we see today have probably been highly adapted through a long period of evolution; they are part of a natural ecosystem." Such stately changes in a grand evolutionary system are exactly what one might expect to produce a rich fauna including many large species of large biomass.

There is a negative historical approach to this same question. If one accepts the view that coextinction is the obverse of coevolution, the extinction patterns also can elucidate the conditions that permit (or do not permit) the evolution of a great chronofauna. Thus, the African savanna-ungulate fauna had been extirpated from the Saharan-Sahelian Region as recently as 8,000 years ago (McIntosh and McIntosh, 1981) by increased aridity. And the fall of the Clarendonian chronofauna can be attributed to similar climatic changes, spread over several million years (Webb, 1977). The details of such linked decimations may shed further light on the conditions that favored the evolution of a rich ungulate fauna.

ACKNOWLEDGEMENTS

From the outset my work on the Clarendonian chronofauna has benefited from the help and encouragement of Richard H. Tedford. Kenneth T. Wilkins gave valuable assistance with the African data. I thank Len Radinsky, Bruce MacFadden, Richard Hulbert, and Ken Wilkins for helpful discussion. My research on Miocene ungulates has been supported by grants

DEB 78-10687 and GB 3862 from the National Science Foundation. This paper is University of Florida Contribution to Vertebrate Paleontology Number 220.

REFERENCES CITED:

BEHRENSMEYER, A.K., and A.P. HILL. 1980. Fossils in the Making: Vertebrate Taphonomy and Paleoecology. Chicago: University of Chicago Press, 338 pp.

BEHRENSMEYER, A.K., and D.E.D. BOAZ. 1980. The Recent Bones of Amboseli National Park, Kenya, in relation to East African Paleoecology; pp. 72-93. *In* Behrensmeyer, A.K., and A.P. Hill (eds.), Fossils in the Making: Vertebrate Taphonomy and Paleoecology. Chicago: University of Chicago Press, 338 pp.

BELL, R.H.V. 1971. A grazing ecosystem in the Serengeti. Scientific American, 224:86-93.

CHANEY, R.W., and M.K. ELIAS. 1936. Late Tertiary floras from the High Plains. Carnegie Institution of Washington Publications, 476:1-46.

CODY, M.L. 1975. Towards a theory of continental species diversities: bird distributions over Mediterranean habitat gradients; pp. 214-257. *In* Cody and Diamond (eds.), Ecology and Evolution of Communities. Cambridge: Harvard University Press.

ELIAS, M.K. 1942. Tertiary prairie grasses and other herbs from the High Plains. Geological Society of America Special Paper, 41:1-176.

GREGORY, J.T. 1971. Speculations on the significance of fossil vertebrates for the Antiquity of the Great Plains of North America. Abh. Hess. Landesamtes Bodenforsch., 60:64-72.

HULBERT, R.C., JR. 1982. Population Ecology of *Neohipparion* (Mammalia: Equidae) from the late Miocene Love Bone Bed of Florida. Paleobiology, 8:159-167.

JANIS, C.M. 1976. The evolutionary strategy of the equidae and the origins of rumen and caecal digestion. Evolution, 30:757-774.

JARMAN, P.J. 1974. The social organization of ungulates in relation to their ecology. Behavior, 48:215-267.

JARMAN, P.J., and A.R.E. SINCLAIR. 1979. Feeding strategy and the pattern of resource partitioning in Ungulates; pp. 130-163. *In* Sinclair, A.R.E., and M. Norton-Griffiths (eds.), Serengeti: Dynamics of an Ecosystem. Chicago: University of Chicago Press.

LAMPREY, H.F. 1963. Ecological Separation of the Large Mammal Species in the Tarangire Game Reserve, Tanganyika. East African Wildlife Journal, 1:63-92.

MACGINITIE, H.D. 1962. The Kilgore Flora: a late Miocene flora from northern Nebraska. University of California Publications in Geological Science, 35:67-158.

MACMAHON, T.A. 1975. Allometry and biomechanics: limb bones of adult ungulates. American Naturalist, 109:547-563.

MATTHEW, W.D. 1901. Fossil Mammals of the Tertiary of northeastern Colorado. American Museum Memoirs, 1:355-447.

MCINTOSH, S.K., and R.J. MCINTOSH. 1981. West African Prehistory. American Scientist, 69:602-613.

OLSON, E.C. This volume. Coevolution or coadaptation? Permo-Carboniferous vertebrate chronofauna. 32 pp.

RADINSKY, L. 1982. Some cautionary notes on making inferences about relative brain size; pp. 29-37. *In* Armstrong, and D. Falk (eds.), Primate Brain Evolution: Methods and Concepts. Plenum Publishing Corporation.

SHOTWELL, J.A. 1958. Intercommunity relationships in Hemphillian (mid-Pliocene) Mammals. Ecology, 39:271-282.

SINCLAIR, A.R.E. 1979. The eruption of the Ruminants; pp. 82-103. *In* Sinclair, A.R.E., and M. Norton-Griffiths (eds.), Serengeti: Dynamics of an Ecosystem. Chicago: University of Chicago Press.

SINCLAIR, A.R.E., and M. NORTON-GRIFFITHS. 1979. *Ibid.* Pp. 1-389.

SKINNER, M.F., S.M. SKINNER, and R.J. GOORIS. 1968. Cenozoic rocks and Faunas of Turtle Butte, Southcentral South Dakota. Bulletin of the American Museum of Natural History, 138: 381-436.

TEDFORD, R.H. 1970. Principles and practices of mammalian geochronology in North America. Proceedings of the North American Paleontological Convention, Part F:666-703.

TEDFORD, R.H., T. GALUSHA, M.F. SKINNER, B.E. TAYLOR, R. FIELDS, J.R. MACDONALD, T.H. PATTON, J.M. RENSBERGER, and D.H. WHISTLER. (In press). Faunal Succession and Biochronology of the Arikareean through Hemphillian interval (late Oligocene through late Miocene Epochs), North America. University of California Publications in Geological Science.

TIESZEN, L.L., and S.K. IMBAMBA. 1980. Photosynthetic systems, carbon isotope discrimination and herbivore selectivity in Kenya. African Journal of Ecology, 18:237-242.

VAN VALEN, L. 1960. A functional index of hypsodonty. Evolution, 14:531-532.

VOORHIES, M.R. 1969. Taphonomy and population dynamics of an early Pliocene vertebrate fauna, Knox County, Nebraska. Wyoming Contributions in Geology, Special Paper no. 1: 1-69. University of Wyoming Press.

VOORHIES, M.R., and J.R. THOMASSON. 1979. Fossil grass anthoecia within Miocene Rhinoceros skeletons: diet in an extinct species. Science, 206:331-333.

WEBB, S.D. 1969. The Burge and Minnechaduza Clarendonian Mammalian faunas of north-central Nebraska. University of California Publications in Geological Science, 78: 1-191.

WEBB, S.D. 1977. A history of savanna vertebrates in the New World. Part 1: North America. Annual Review of Ecological System, 8:355-380.

WEBB, S.D., B.J. MACFADDEN, and J.A. BASKIN. 1981. Geology and paleontology of the Love Bone Bed from the late Miocene of Florida. American Journal of Science, 281: 513-544.

WESTERN, D. 1979. Size, life history and ecology in mammals. African Journal of Ecology, 17:185-204.

WESTERN, D. 1980. Linking the ecology of past and present mammal communities; pp. 41-54. *In* Behrensmeyer, A.K., and A.P. Hill (eds.), Fossils in the Making. Chicago: University of Chicago Press.

WOLFE, J.A. 1978. A paleobotanical interpretation of Tertiary climates in the northern Hemisphere. American Scientist, 66:694-703.

WOOD, H.E., II, R.W. CHANEY, J. CLARK, E.H. COLBERT, G.L. JEPSEN, J.B. REESIDE, JR., and C. STOCK. 1941. Nomenclature and correlation of the North American continental Tertiary. Bulletin of Geological Society of America, 52: 1-48.

COEVOLUTION OR COADAPTATION? PERMO-CARBONIFEROUS VERTEBRATE CHRONOFAUNA

Everett C. Olson

Department of Biology
University of California, Los Angeles
Los Angeles, California

A well-defined vertebrate community (chronofauna) has been followed through moderately continuous sedimentary deposition over a period of approximately 15 million years. The fossils deposited during the last 5 to 7 million years are given special attention. The deposits form the beds of the uppermost Carboniferous and Lower Permian in Oklahoma and Texas. Persistent associations and concurrent changes of members of the vertebrate lineages, species lineages, occur. The nature of the associations and concurrent changes are analyzed. The record provides the following criteria for interpretive purposes: (1) geological, temporal and ecological associations of the entities; (2) morphological aptness for interactions; (3) concurrence of morphological changes in associated lineages; (4) direct evidences of interactions.

Each of these is examined in the evolving community. Outcomes of the examination are viewed over the spectrum of patterns of changes induced by selection dominated by direct interactions between the members of the lineages to those induced by selection dominated by independent reactions of lineage members to their common environmental milieu. The role of the top predator, Dimetrodon, *in the changing system is singled out as the focal point for the study. Coadaptation of lineage members to each other* and *to changes in response to modifying physical environment, emerges as the favored hypothesis. The hypothesis has predictive value and is subject to further testing. Coevolution may be inferred but the data are insufficiently sensitive and precise to provide for meaningful tests.*

INTRODUCTION

A coherent assemblage of fish, amphibians and reptiles existed during the time of deposition of the Clear Fork group and its equivalents (Lower Permian) in Texas and Oklahoma (fig. 1). Remains of this assemblage are preserved in the red beds exposed in a band running roughly north-south for about 300 miles and ranging from about 300 to 1600 feet in thickness and reaching a maximum thickness in North Central Texas. The members of the assemblages were described as

	Group	Formation
LOWER PERMIAN	CLEAR FORK	Choza
		Vale
		Arroyo
	WICHITA	Lueders*
		Clyde*
		Belle Plains
		Admiral
		Putnam
		Moran
		Pueblo

*Figure 1. The Lower Permian section of Texas. *Referred to Clear Fork Group in some publications, but here grouped with Wichita formations to indicate close faunal affinities with Admiral and Belle Plains formations.*

forming an integrated, evolving system termed a chronofauna by the writer (Olson, 1952a). The structure of the system was described from the perspective of a trophic complex with the top predator, *Dimetrodon*, at the apex of the food pyramid. This system provides an opportunity to explore the modes of interaction of its constituents and the changes in interactions over a period of five to seven million years. The central question of this paper is the extent to which these interrelationships represent either coevolution or coadaptation, when taken as ends of a spectrum of interactions, and the extent to which the materials permit testable conclusions.

Emphasis is placed upon the Clear Fork assemblage, but its interpretation requires information on the faunal assemblages of the pre-Clear Fork Wichita deposits of Texas and Oklahoma. Although the faunas preserved in the beds of the Wichita formations have been collected for many years, the work (except by Case, e.g., Case, 1915, 1919) was not carried out with ecological interpretations in mind. As the result the Wichita samples are not, for the most part, reliable for the taphonomic analyses basic to an understanding of the living populations that they represent.

The study area of this paper is limited geographically to deposits found in Texas and Oklahoma. Somewhat similar collections have come from New Mexico, Colorado, Arizona and the Ohio-Pennsylvanian-West Virginian areas. They are age equivalents of the collections from formations of the Wichita Group of Texas and equivalents in Oklahoma. Collections covering the full Lower Permian obtained in Europe are sufficiently remote geographically that they can be reasonably excluded from this study.

The ecological setting of the faunas of the Lower Permian of Texas and Oklahoma was broadly a near ocean, terrestrial environment with a tropical to subtropical climate and a reduced Coal Measures type flora. During deposition of the beds of the Wichita Group high humidity and ample rainfall persisted throughout the year. Drying trends and increasing seasonality began with the onset of the Clear Fork and persisted, intensifying, until the end of the Lower Permian (see e.g., Olson, 1958).

The topography ranged from very low relief near the ocean margins to moderate relief inland. The shoreline fluctuated, continuing the cyclical mode of deposition of the Pennsylvanian. Following deposition of the Lueders limestone (fig. 1) fluctuations were reduced, waning up to the time of development of evaporite basins near the end of the Lower Permian.

The vertebrates, along with remains of terrestrial plants and fresh water invertebrates, were deposited in a more or less continuous sequence of beds over a period of about 15 million years. Fossils are preserved in sediments formed in bodies of standing water, here termed ponds, in swamps, under nearshore circumstances, in river and stream beds and in overbank and flood plain areas. Vertebrate remains, including some terrestrial animals, occur in some of the limestone deposited during the periodic transgressions of the seas.

METHODS

The chronofaunal model

The basis of this study is the chronofaunal model diagrammed in fig. 2. Trophic relationships form the basis for this model and represent the flow of energy through the system. The concept is one of a stable network of organisms, the vertebrate community, each element of which plays a

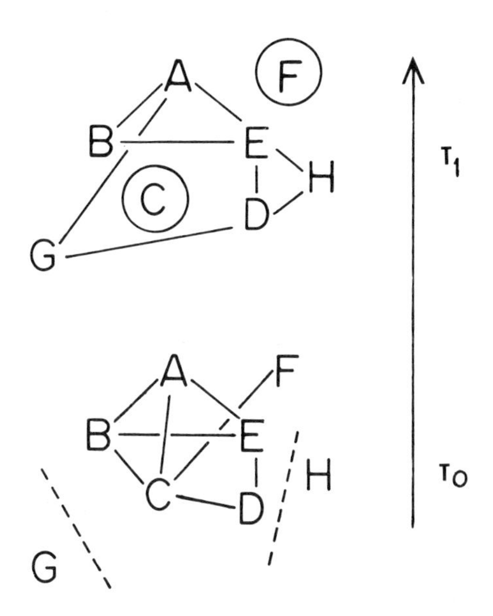

Figure 2. Diagram of the chronofaunal model. A through G, species (genera) each of which plays a particular role in the association of the elements of the system in relation to other elements and the physical environment. Solid lines indicate principal trophic relationships of elements, with the direction of action unspecified. t_0 *= observed base system,* t_1 *= a later time. At* t_1*, F has dropped from the system and its role no longer exists; G has replaced C, which is eliminated from the system, and G has assumed the role of C; H has been added with a trophic relationship to E and D, adding a new role to the system.*

particular role in the community structure. The network tends to remain the same throughout the duration of the system, being modified, if at all, by:

(1) elimination of one or more positions;

(2) change in a species occupying a position;

(3) addition of one or more structural element, either from within or without the system.

This model is simple and certainly simplistic, but provides a useful framework for the study of evolution of a community and the behavior of its constituents.

The range of interactions

Interactions between species must be inferential and can at best be crudely resolved in most cases. Generic rather than specific groups are the most appropriate for study of ecological roles in the Permian vertebrate system and this level is used except where species relationships become ecologically important and resolvable. The following criteria provide the basis for interpretations of interactions and are treated in detail in sequence in the following pages:

(1) The associations of taxa in samples. Inferences about the ecological meaning of associations require taphonomic analysis. The likelihood of correct interpretations of associations as representative of those in the living populations increases with an increase in the number of samples.

(2) The morphology of assumed associates. Morphologies must be demonstrably appropriate for the inferred trophic relationships and for the environments in which the organisms are thought to have lived.

(3) Direct evidence such as the remains of prey in the enterospira or coprolites of predators, tracks, trails, marks on bones or associations in regurgitated residues.

(4) The extent to which the morphologies in inferred associates remain congruent through time.

All organisms in an ecological system interact to some extent. Interactions in an interspecies population range from mere use of the common life sustaining resources, such as oxygen and water, to levels of symbiosis from commensalism to mutualism. Translated into evolutionary terms, mutualism is expressed as coevolution in which interactions between a pair of taxa, usually species, provide the principal selective factors (fig. 3). At the other end of the spectrum coadaptation involves the congruent changes of members of a species pair in which the dominant selective factors are those of the environment, which act independently on each of the members. The extent to which such relationships can be deduced from paleontological and sedimentological evidence is a primary problem in interpretations of ecological evolution. The same principles apply in all cases, but the extent to which valid interpretations can be made varies widely under different circumstances. Each case requires its own analysis and

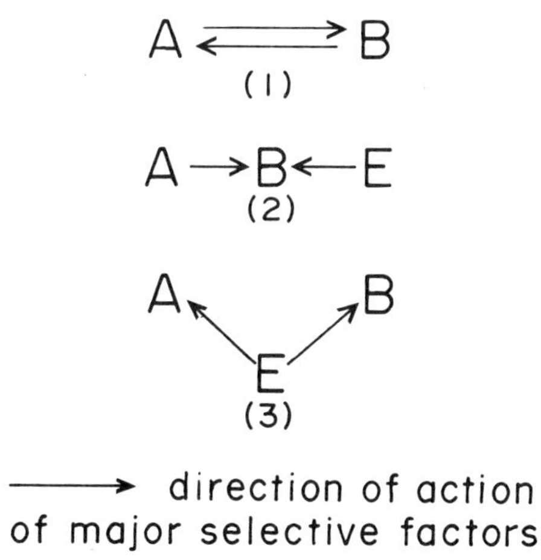

Figure 3. Diagrams of the interaction patterns of a species pair. The arrows point in the direction of action, from source to recipient, of the principal selective factors during evolutionary change. A and B are species. E indicates the environment of the species excluding the reciprocal interaction.

results can be extended only to instances which can be shown to be similar in major parameters.

MATERIALS AND DATA ON INTERACTIONS

Associations

Much of the general information on the genera used in this study is shown in tables 1 through 4 and figures 4, 5, and 6. Well over 200 samples from the Wichita and Clear Fork have furnished the data. The most definitive data are from the Clear Fork and come from about 100 samples collected over a stretch of about 300 miles and a section up to 1600 feet thick.

Many more than the 18 genera listed in figure 4 are known from this sequence, but only those that can be shown to be ecologically significant in our analyses are used. These may be divided into the following categories:

(1) Organisms that played a major role in the food chain, but were not primary prey of *Dimetrodon*, including *Xenacanthus*, paleoniscoid genera, *Sagenodus*, *Gnathorhiza*, *Lysorophus*. All were aquatic and fed variously on other organisms, both plants and animals, thus providing a vertebrate food base for the prey animals of *Dimetrodon*.

(2) The "large" prey animals of *Dimetrodon* with estimated weights ≤ 40 kg, including *Eryops*, *Diadectes*, *Edaphosaurus*, *Ophiacodon*. They range from semi-aquatic to terrestrial.

(3) The "small" prey animals of *Dimetrodon*, with estimated weights of ≥ 10 kg, including *Diplocaulus*, *Archeria* (including *Cricotus*),

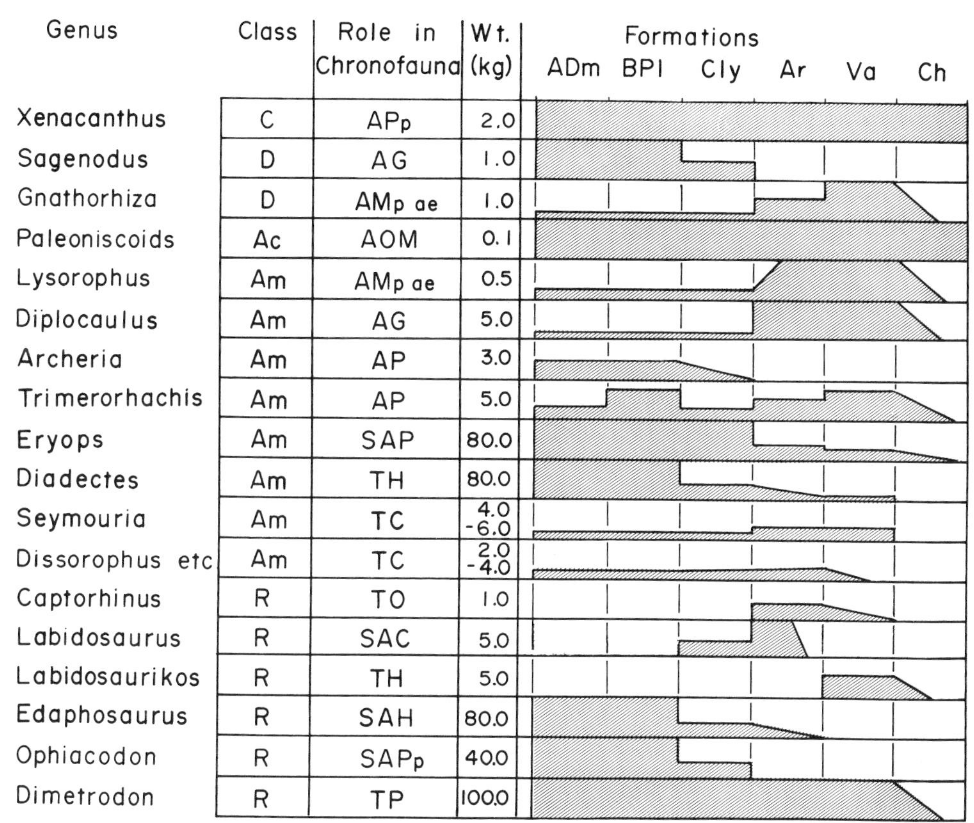

Genus	Class	Role in Chronofauna	Wt. (kg)
Xenacanthus	C	APp	2.0
Sagenodus	D	AG	1.0
Gnathorhiza	D	AMp ae	1.0
Paleoniscoids	Ac	AOM	0.1
Lysorophus	Am	AMp ae	0.5
Diplocaulus	Am	AG	5.0
Archeria	Am	AP	3.0
Trimerorhachis	Am	AP	5.0
Eryops	Am	SAP	80.0
Diadectes	Am	TH	80.0
Seymouria	Am	TC	4.0 -6.0
Dissorophus etc	Am	TC	2.0 -4.0
Captorhinus	R	TO	1.0
Labidosaurus	R	SAC	5.0
Labidosaurikos	R	TH	5.0
Edaphosaurus	R	SAH	80.0
Ophiacodon	R	SAPp	40.0
Dimetrodon	R	TP	100.0

Figure 4. Seventeen genera and the paleoniscoid suite of genera that are important in the Permo-Carboniferous chronofauna complex. Abbreviations: Classes--*AC, Actinopterygia; AM, Amphibia; C, Chondrichthyes; D, Dipnoi; R, Reptilia.* Roles--*ae, aestivators; AG, aquatic "grubbers"; AMp, aquatic micropredators; AOM, aquatic feeders on various small foodstuffs; AP, aquatic predators; APp, aquatic piscivorous predators; SAC, semiaquatic carnivores; SAH, semiaquatic herbivores; SAP, semiaquatic predators; SAPp, piscivorous semiaquatic predators; TC, terrestrial carnivores; TH, terrestrial herbivores; TO, terrestrial omnivores; TP, top predator.* Formations--*ADm, Admiral; Ar, Arroyo; BPl, Belle Plains; Ch, Choza; CLy, Clyde and Lueders; Va, Vale.*

Trimerorhachis, *Seymouria*, *Captorhinus*, *Labidosaurus*, *Captorhinikos*, *Labidosaurikos*. Of these, *Seymouria* and the captorhinids, excepting *Labidosaurus*, were predominantly terrestrial.

(4) *Dimetrodon*, the top predator with an average estimated weight of 100 kg.

Only the members of the last three groups enter into the detailed analyses, but the importance of the underlying food base is tacit in the discussions. Special attention will focus on the close association of *Dimetrodon* and *Diplocaulus* later in the paper. For a more general analysis six samples selected from standing water deposits from the Wichita and Clear Fork are used. Accumulations of vertebrates in such deposits would appear to provide the best chance of procurement of samples from living populations deposited more or less *in situ*. The six samples are listed in table 1. Sample sizes range from 30 to 316 individuals, based on number of identified specimens in the sample. Specimens are for the most part fragmentary and the chances of overestimation of numbers of some components are high. Throughout the Clear Fork the proportions of individuals are fairly well matched by samples from other pond deposits, but Wichita samples show greater variability (see Olson, 1975).

Figure 5 is a plot of the percentages of large and small prey animals of *Dimetrodon*. Two features of the diagram are particularly significant. First is the sharp drop in percentages of large prey and the increase in percentages of small prey between the time of deposition of the Clyde and middle Arroyo, especially as shown in 5B. This must have had a direct impact on the predatory habits of *Dimetrodon*. Second is the consistently high percentage of *Dimetrodon*.

Table 1. *Numbers and percentages of individuals of genera in each of 6 pond samples. Data for Admiral and Belle Plains from Romer, 1928; for Clyde and middle Arroyo from Olson field notes; for lower Vale from Olson, 1958, augmented; for middle Vale, Olson and Mead, 1982, augmented.*
Abbreviations: ADm, Admiral Formation, Briar Creek, Archer County, Texas; BPl, Belle Plains Formation, south of Fulda, Baylor County, Texas; CLy, Clyde and Lueders Formations, west Mitchel Creek, Baylor County, Texas; LVa, lower Vale Formation, western Knox County, Texas; MAr, middle Arroyo Formation, Pond Creek locality, Grant County, Oklahoma; MVa, middle Vale Formation, 6 miles southeast of Haskell, Haskell County, Texas. N, number of individual specimens.

	ADm		BPl		CLy		MAr		LVa		MVa	
	N	%	N	%	N	%	N	%	N	%	N	%
Eryops	56	18	37	28	6	19	4	4	2	7	7	6
Diadectes	3	1	18	14	2	6	6	6	0	0	0	0
Ophiacodon	25	8	8	2	2	6	0	0	0	0	0	0
Edaphosaurus	21	6	4	3	2	6	2	2	0	0	0	0
Trimerorhachis	3	1	30	23	4	13	10	10	2	6	2	2
Archeria + Cricotus	43	14	0	0	1	3	0	0	0	0	0	0
Diplocaulus	1	--	0	0	1	3	42	42	15	50	58	50
Captorhinomorpha	0	0	0	0	3	10	3	3	1	3	3	3
Seymouria	0	0	0	0	0	0	2	2	0	0	10	8
Dimetrodon	164	52	38	29	10	32	32	32	10	33	35	30
TOTALS	316	100	130	99	31	98	101	101	30	99	117	101

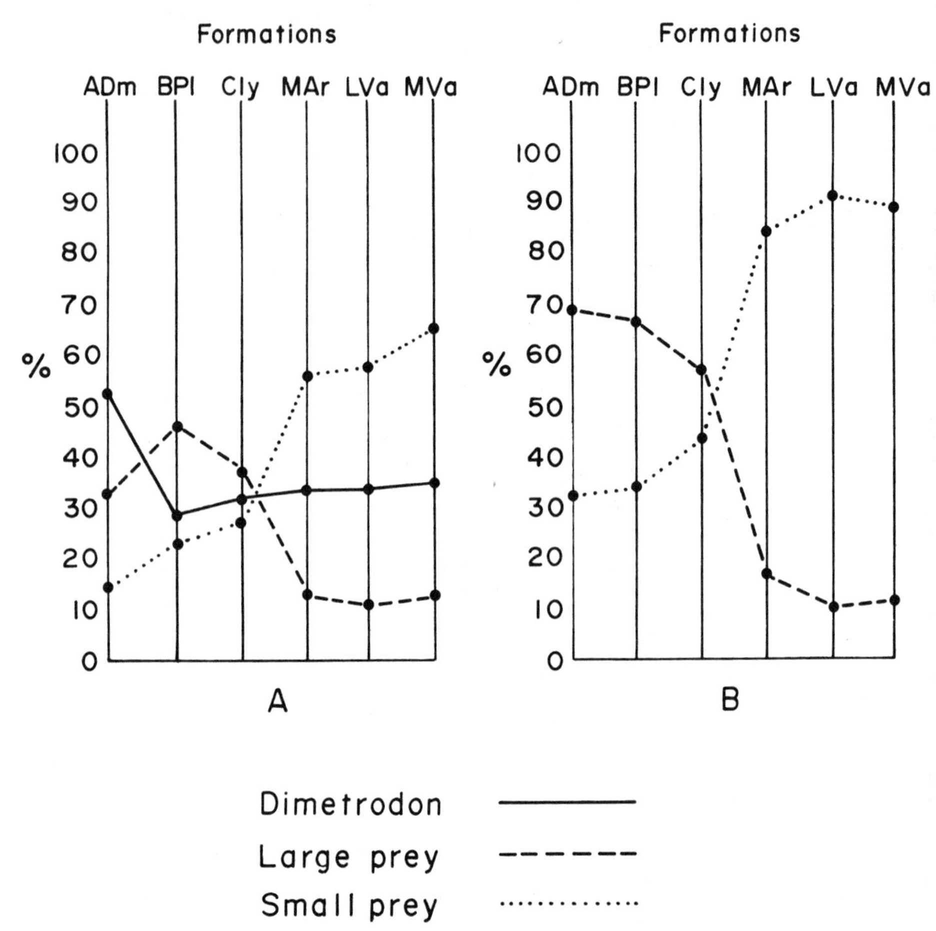

Figure 5. Six pond samples plotted from data in table 1. Percentages of large and small prey animals of Dimetrodon *based on number of individuals. A, percentages based on samples with* Dimetrodon *included; B, based on percentages with* Dimetrodon *excluded. Abbreviations as in table 1.*

Not evident in figure 5 is the relative contributions to the total biomass of the various components, tables 2 and 3. Figure 6 shows estimates of biomass based on weights of genera

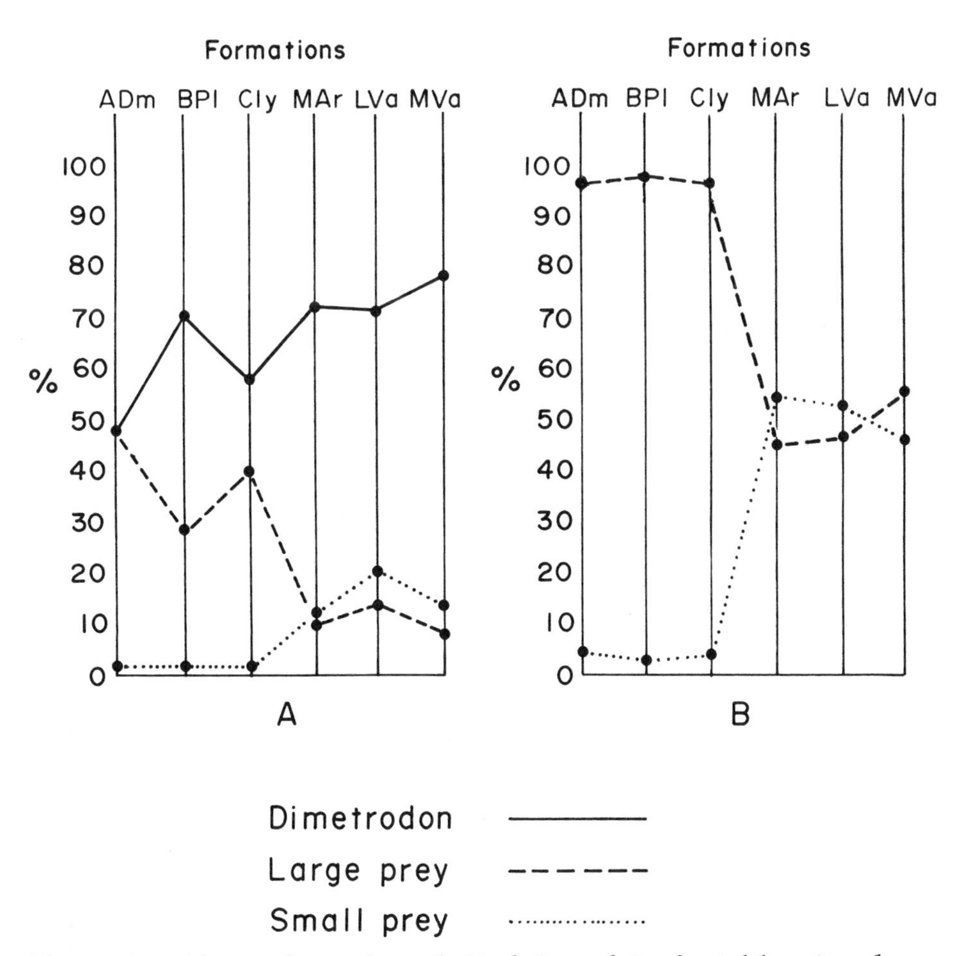

Figure 6. Six pond samples plotted from data in tables 2 and 3 based on biomass estimations. A, based on percentages with Dimetrodon *included; B, based on percentages with* Dimetrodon *excluded.*

as given in figure 4. These estimates, although very crude, reveal the same drop in dominance of the large prey animals from the Clyde to the Arroyo, but also indicate a greater continuing significance of large prey.

Table 2. Amounts and percentages of estimated biomass of genera of each of the 6 pond samples. Captorhinidae genera lumped together. In second column of each formation the value to the left is for the sample with Dimetrodon *excluded and that to the right with* Dimetrodon *included. For abbreviations see table 1.*

	ADm		*BPl*		*CLy*		*MAr*		*LVa*		*MVa*	
	Bi	*%*	*Bi*	*%*	*Bi*	*%*	*Bi*	*%*	*Bi*	*%*	*Bi*	*%*
Eryops	1850	48,24	2800	39,12	500	65,28	100	17, 5	100	24, 7	350	38, 8
Diadectes	1440	37,19	1440	20, 6	160	21, 9	80	14,44	100	24, 7	160	17, 4
Ophiacodon	120	3, 1	1000	14, 4	80	10, 5	0	0, 0	0	0, 0	0	0, 0
Edaphosaurus	320	8, 4	1680	24, 7	0	0, 0	80	14, 4	0	0, 0	0	0, 0
Trimerorhachis	150	4, 2	15	-, -	20	3, -	50	9, 2	10	2, 1	0	0, 0
Archeria-Cricotus	0	0, 0	177	2, -	3	-, -	0	0, 0	0	0, 0	0	0, 0
Diplocaulus	0	0, 0	0	0, 0	0	0, 0	252	44,12	180	44,13	348	38, 8
Captorhinidae	0	0, 0	0	0, 0	3	-, -	3	-, -	6	1, -	15	2, -
Seymouria	0	0, 0	0	0, 0	0	0, 0	8	1, -	15	4, 1	48	5, 1
Dimetrodon	370	0,48	16400	0,70	1000	0,58	1500	0,72	1000	0,71	3500	0,78
TOTAL including Dimetrodon	7850		23512		1766		2073		1411		4421	
TOTAL less Dimetrodon	3880		7112		766		573		411		921	

Table 3. Summation of biomass percentages for large and small prey and Dimetrodon. *Upper section with* Dimetrodon *included, lower with* Dimetrodon *excluded. For abbreviations see table 1.*

	ADm	*BPl*	*CLy*	*MAr*	*LVa*	*MVa*
Including Dimetrodon						
Large prey	48	29	42	13	14	12
Small prey	2	1	0	14	15	9
Dimetrodon	48	70	58	72	71	78
Excluding Dimetrodon						
Large prey	96	98	96	45	48	55
Small prey	4	2	3	55	52	45

The dominance of *Dimetrodon* based on biomass far exceeds that indicated by numbers of individuals. As far as these samples are concerned a major discrepancy exists both in numbers and mass of the top predator relative to the potential prey animals. Irrespective of the trophic relationships that may be hypothesized, *Dimetrodon* is certainly immensely over-estimated. This applies in general to pond deposits, not just the six included in this study. In part this results from sample bias and ease of recognition, for *Dimetrodon* is large, its parts are readily recognized and its long neural spines produce large numbers of fragments which can be readily assigned to the genus. Even in cases where minimum numbers can be obtained, however, the proportion of *Dimetrodon* remains very high (Olson and Mead, 1982).

Two excellent samples showing relatively high percentages of *Dimetrodon* are the Arroyo Craddock bone bed (Williston, 1911) and the lower Vale, Sid McAdams bone bed (Olson and

Mead, 1982). They have been the basis of an interpretation of the ponds in which such accumulations took place as "carnivore traps." The high concentration of *Dimetrodon* may have resulted from concentrations of prey in and about the ponds, but it still leaves open the question of why *Dimetrodon* seems to have undergone mass deaths at the feeding sites. There remain many unanswered questions. The enigma is complicated by the fact that some drying ponds, such as one in the Wellington formation (Wichita equivalent) near Orlando, Oklahoma, lacks a concentration of *Dimetrodon*, even though the genus existed in the area (Olson, 1975).

Morphologies and Functions

A compatibility of morphologies and the functional relationships of the constituents are implicit in the chronofaunal concept. Briefly the morphologies of the critical members of the Permo-Carboniferous system relative to their roles are as follows.

Dimetrodon (fig. 7). The skeleton and dental morphology of *Dimetrodon* are suited only to a carnivorous mode of life and strongly suggest a predatory existence (see e.g., Barghusen, 1973). Its primary locomotor and feeding mechanisms were well suited to capturing large prey and rending pieces of flesh from the skeleton, to be swallowed with little mastication. This mode of feeding is highly compatible with its role during the Wichita, but strong behavioral modifications must have occurred with the onset of the Arroyo.

Ten species, some tentative, were identified by Romer and Price (1940). This number probably is excessive (e.g., Olson and Mead, 1982). Differentiation was based on overall size, the form of the posterior part of the skull, the shape

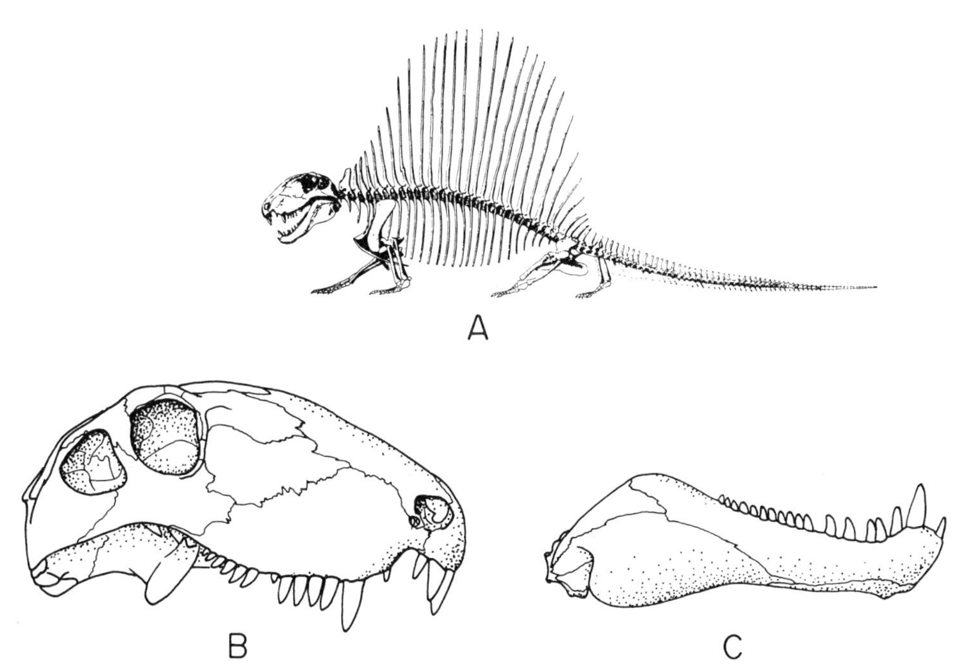

Figure 7. Dimetrodon. *A, skeleton; B, skull; and C, lower jaw. After Romer and Price, 1940.*

of the long neural spines and relative lengths of the vertebral centra. Except for size, the differences are minor and morphologically insignificant as far as the functions of *Dimetrodon* as a predator are concerned. The weight estimate used in this paper is 100 kg, which seems high, but is less than that for several species made by Romer and Price (1940). The 100 kg estimate is an average and the estimates of Romer and Price were made on large, fully mature individuals. If, however, the 100 kg estimate is low, the problem of excess biomass of *Dimetrodon* is exacerbated.

Large prey animals. Eryops has a stocky body, a long heavy tail, short massive limbs and a proportionately very large skull. Its marginal teeth were sharp, labyrinthine cones. The morphology indicates that *Eryops* was primarily at home in the water, capable of moving slowly but not of feeding on land. It was a carnivore adapted to feeding on small to moderate sized aquatic animals, some of which may have been drawn into its massive oral cavity as the gape of the mouth was rapidly enlarged. Among the larger, more active prey foods was the shark, *Xenacanthus*, remains of which have been found in *Eryops* coprolites.

Diadectes had a massive skeleton marked by broadly expanded neural arches and overlapping ribs. Its limbs, although massive, were proportionately longer than those of *Eryops*. It was certainly at home on land. The skull and jaws were formed of very thick, spongy bone and were small relative to body size. The dentition consisted of long, peg-like upper "incisors" and transversely widened, tuberculated cheek teeth.

Ophiacodon was characterized by a long, lightly built skull and lower jaws. The upper and lower jaws are set with long rows of sharp, conical teeth characteristic of piscivorous animals. The postcranium is light and seemingly rather small in proportion to the skull. The animal was capable of living on land but both its skeleton and dentition indicate that water was its primary habitat.

Edaphosaurus had a very small skull, short, heavy jaws, blunt marginal teeth and a pavement of pebbly palatal teeth. These all indicate that *Edaphosaurus* was an herbivore but its food has not been precisely determined. It occurs, as a rule, in deposits that suggest an aquatic habitat. The postcranium, best known from the long neural spines with cross-bars, ranged from lightly structured in the Wichita species to robust in the Arroyo species. Its limbs, while massive, were well suited to locomotion on land. If not actually semi-aquatic

Edaphosaurus clearly lived near the water.

Each of these animals could have provided ready prey for *Dimetrodon* which was active and well equipped to subdue and kill them and to tear flesh from the skeletons. Their frequent association with *Dimetrodon* both in pond and stream deposits suggests that this could have been the nature of their relationship, whereas lack of associations would preclude such an interpretation.

Small prey animals. The skulls and jaws of two genera, *Trimerorhachis* and *Archeria* were set with long arrays of sharp conic teeth. The skeletons were elongated and the limbs proportionately very small. These structures indicate active, predaceous swimming animals. The morphology conforms with their persistent occurrence in deposits formed in standing water and streams. *Trimerorhachis*, in particular, had the ability to move about on land, but the very small limbs and a somewhat fish-like tail suggest that water was the primary habitat. *Trimerorhachis* persisted as a common constituent through most of the time of Clear Fork deposition, but *Archeria* did not survive the Clyde.

Diplocaulus (figs. 8 and 9) was a flat bodied, very short limbed animal with a broad, boomerang-shaped skull and weak jaws set with small teeth. It was a strictly aquatic animal, possibly an aestivator during its younger stages. Its food probably consisted of small invertebrates, small vertebrates, plants and perhaps organic debris.

The other small prey animals were primarily terrestrial and major constituents only of Clear Fork samples. *Seymouria* has the sharp teeth of a carnivore, a relatively heavy postcranium and moderately short limbs. A small species existed during deposition of the Arroyo Formation and a larger one during the Vale. The captorhinids were more lightly built,

Figure 8. Diplocaulus, *a reconstruction by Dr. Kathryn Bolles.*

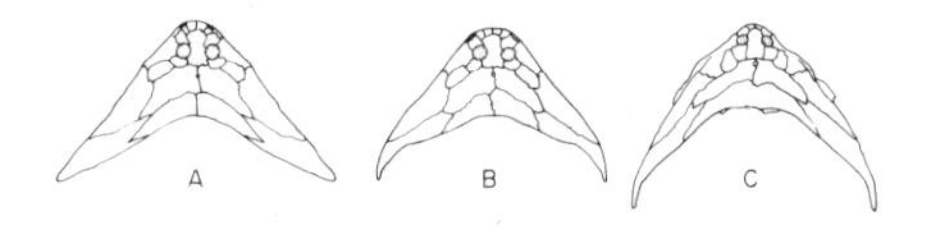

Figure 9. Skulls of Diplocaulus. *A,* D. magnicornis; *B,* D. recurvatus; *C,* D. brivirostris. *A, modified after Douthitt, 1917; B, after Olson, 1952b; C, after Olson, 1951.*

with the relative robustness of the skeletons increasing in the larger genera. Limbs and girdles suggest that the smaller representatives were active, somewhat lizard-like terrestrial animals. Marginal teeth were set in single rows in the Wichita genera but formed multiple rows in Clear Fork genera except for *Labidosaurus*, a moderately large, semiaquatic genus of the Arroyo. The early genera appear to have been carnivores and insectivores but during the Clear Fork a dietary shift took place in the direction of herbivory. This shift was accompanied by general size increase, leading to *Labidosaurikos* of the Vale and Choza with an estimated weight of 6 kg in contrast to the 1-2 kg estimate of *Captorhinus* of the Arroyo.

Finally, in addition to the genera noted above and entered in the tables, a number of other possible prey animals of small to moderate size existed, especially during the time of deposition of the Arroyo beds. Prominent among these were members of the Dissorophidae. They were primarily carnivores with skulls and jaws set with small, sharp, conical teeth. The postcrania, often with a partial dorsal armor, were lightly built and the limbs were long and slender. The structure is commensurate with their usual modes of occurrence, which suggest a terrestrial habitat. They form additional possible terrestrial prey for *Dimetrodon*, but are too sporadic in occurrence and associations to enter into the calculations in this paper.

Direct Evidence

Direct evidence of associations is uncommon in the Permian red beds. Coprolites and enterospira, which will be lumped as coprolites for convenience, are abundant but pose two problems. One is the difficulty of identification of the producer of the coprolite. The other is that the slow digestive processes of ectotherms generally assure that only the most resistant structures will be preserved. As a rule fish scales and occasionally teeth are the only identifiable specimens in the coprolites of predators.

The coprolites of *Dimetrodon* can, in some instances, be recognized with confidence. They contain mascerated bone, indicating the producer was a carnivore, and many coprolites are too large to have been the product of any other common carnivore of the complex. When undistorted they are torpedo-shaped with one end round and blunt and the other tapered. In rare instances unaltered bone fragments occur in these coprolites and some of these have been identified as vertebral and skull parts of *Diplocaulus*.

Coprolites associated with the remains of *Eryops* are more symmetrical in outline than those attributed to *Dimetrodon*. One specimen, from a large collection found closely associated with specimens of *Eryops*, contains a part of a chondrocranium of the shark *Xenacanthus*. The most commonly identifiable coprolites and enterospira are those of this particular shark (Williams, 1972). These usually contain abundant scales of paleoniscoid fishes.

Little other direct evidence is sufficiently trustworthy to be useful. Cases of skulls pierced by teeth have been found, bones appear to have been cut by the teeth of predators or scavengers and concentrations of regurgitated bones have been found. In none of these cases is the identity of the responsible carnivore known.

Congruent Changes of Morphologies of Associates

The principal anatomical structures of *Dimetrodon* related to capture and ingestion of prey do not change through the Wichita-Clear Fork sequence except for the size differences of the supposedly different species. *Diadectes* and *Eryops* of the large prey animals, likewise show no appreciable changes. *Edaphosaurus*, however, shows an increase in robustness and a large Clyde species of *Ophiacodon* marks a similar change. In neither of these cases do the changes in size and proportions seem sufficient to challenge the predatory capacities of adult individuals of *Dimetrodon*.

Among the smaller prey, the aquatic *Trimerorhachis insignis*, shows no changes and two poorly known species that arose after the onset of the Clear Fork differed only in minor features such as the position of the orbits and the depths of the skull. *Archeria* shows no significant change, but did not persist into the Clear Fork. Important changes did occur in *Seymouria* and in the captorhinids but only after the beginning of the Clear Fork deposition.

The significant changes, thus, took place during the deposition of the Clear Fork formations, and these are correlated with the onset of drying. The sharp drop in the relative number or biomass of the large prey animals and numerical increase in small prey animals as diagrammed in figures 5 and 6 has been stressed earlier. Its impact on *Dimetrodon* should, it would seem, have been considerable. To look into this further we will concentrate on the smaller prey animals, for those large prey animals that did persist into the Clear Fork in the area studied either became extinct soon or remained in much reduced numbers.

The persistence of *Trimerorhachis insignis* in considerable numbers has been noted. Its role remained unchanged, except that it formed a numerically larger part of the prey sample during the Clear Fork. *T. mesops* occurs in the Arroyo and Vale (Olson, 1955; Olson and Mead, 1982). *T. rogersi*, of the Choza, developed from *T. insignis*. Both were about the same size as *T. insignis* and like that species lived in the water. The significance of the speciation is not known, but it does not seem to be related to the prey-predator relationship of *Dimetrodon* and poses no evident basis for any change in *Dimetrodon*.

Diplocaulus increased greatly in numbers during the first phase of the Clear Fork deposition and maintained high numbers until the upper part of the Choza. During the transition from the Arroyo to the Vale *D. magnicornis* gave rise to *D. recurvatus*, an event which accompanied a shift from pond to stream environment. *D. recurvatus* ecologically replaced an earlier (Arroyo) species, *D. brevirostris*, which had become extinct soon after the beginning of Vale deposition. The morphological changes between *D. magnicornis* and *D. recurvatus* were slight in essence involving a convergence of the latter with *D. brevirostris* in the posterior deflection of the tabular "horns" (fig. 8).

Dimetrodon and *Diplocaulus* are the most constantly associated genera in the Clear Fork. As shown in table 4, in 50 sites in which one or the other or both genera are found the two occur together in 38 of those, or 76 percent. If flood plain deposits, formed under circumstances unsuitable for *Diplocaulus*, are omitted, the percentage rises to 92 percent. Both genera appear to have maintained an adaptation to increasingly dry climates, at least until the end of the Vale. Thereafter, with the beginning of evaporite conditions of the Choza, samples are small and scattered and unreliable as data on abundances and associations. During the Clear Fork the morphology of *Dimetrodon* remained unchanged whereas that of its principal associate was modified, as treated further in the following section.

Both the captorhinids and *Seymouria* underwent increases

Table 4. Associations of Dimetrodon *and* Diplocaulus *in Clear Fork samples. Data on Arroyo from field notes of Olson; on Vale and Choza from Olson, 1958, slightly modified by additional collecting. N, number of samples; 1, number of samples with both* Dimetrodon *and* Diplocaulus*; 2, number of samples with only* Dimetrodon *or* Diplocaulus.

	Arroyo			*Vale*			*Choza*		
	N	*1*	*2*	*N*	*1*	*2*	*N*	*1*	*2*
Pond	12	12	0	7	7	0	1	0	1
Stream	2	0	2	17	17	0	1	1	0
Flood plain	3	1	2	5	0	5	2	0	2
TOTALS	17	13	4	29	24	5	4	1	3

in size and modifications of morphological structures during the Clear Fork. This was accompanied by an increase in relative numbers, foreshadowed in the Clyde among the captorhinids. *Captorhinus*, the first known genus with multiple rows of maxillary and mandibular teeth, first appeared in the base of the Arroyo. It was a small genus with estimated weight ranging from less than a kilogram to perhaps 2 kg. Accompanying it was a larger genus, *Labidosaurus*, about 4 kg in estimated weight. *Labidosaurus* did not persist into the upper part of the Arroyo. *Labidosaurikos* was a larger genus which appeared in the basal Vale and was a moderately common constituent of samples from the Vale and Choza. Its jaws were set with several rows of regularly arranged teeth. With it was a smaller genus, *Captorhinikos*, which was morphologically somewhat intermediate between *Captorhinus* and *Labidosaurikos*. The small species of *Seymouria*, *S. baylorensis*, of the Arroyo and basal Vale gave way to the larger *S. grandis* during the middle and upper Clear Fork.

Both the captorhinids and *Seymouria* occur in association with *Dimetrodon* and were suitable terrestrial prey. To a minor degree they may have assumed the ecological roles of such large terrestrial animals as *Diadectes*, which decreased sharply over the course of Clear Fork deposition.

EVOLUTIONARY INFERENCES

If the data and interpretations of associations and trophic relationships of the genera of the Permo-Carboniferous vertebrate chronofauna are taken as broadly valid, the evolutionary inferences discussed in the following paragraphs can be made. The Wichita-Clear Fork vertebrate complex shows a strong response to changes in physical environment, resulting in a very differently balanced trophic web as the wet climates of the Wichita gave way to the progressively drier climates of the Clear Fork.

Similar climatic modifications took place over all of the circumequatorial regions of Pangea for which there is a non-marine record for the later part of the Lower Permian and the Upper Permian. In the areas most proximate to the mid-continent of North America the onset of drier climates began somewhat earlier than in the Texas-Oklahoma region (Vaughn, 1969a,b; Olson and Vaughn, 1970). There is no clear evidence that the accompanying faunal changes, best known in New Mexico and Colorado, had a direct impact on the faunas of the mid-continent area. It is thus assumed that the faunal changes observed in the Texas-Oklahoma area were the direct result on interactions of the indigenous faunas and the environmental changes, not the result of influxes of organisms from other regions.

Only a moderate number of easily recognized morphological

modifications took place among the prominent constituents of the Wichita-Clear Fork assemblage, the most evident being those of *Diplocaulus*, *Seymouria* and the captorhinids. Minor or obscure evolutionary changes may have occurred in these and other genera but, whatever these may have been, they appear to have been ecologically insignificant and not to have altered the pair relationships between *Dimetrodon* and the other animals of the trophic web.

The principal changes that took place during the time represented by deposition of the formations of the Wichita and Clear Fork may be summarized as follows:

(1) losses of some genera from the system;
(2) entrance of new genera and species into the system;
(3) strong shifts in relative abundances of genera through time;
(4) evolutionary changes in some groups during the Clear Fork.

Behavioral and physiological adjustments to changes in the physical environment presumably took place among the lineages that persisted from the Wichita into the Clear Fork. Most successful in accomplishing such adjustments were *Trimerorhachis* and *Dimetrodon*, which remained abundant and morphologically unchanged. The former persisted in aquatic environments similar to those of the Wichita and presumably needed to undergo little adjustment. The latter, however, had to adjust not only to increasingingly xeric conditions but to a sharp change in its principal prey. The latter adjustment involved both the marked change in prey with the onset of the Clear Fork and the continuing changes in habitats and morphology of prey during the Clear Fork. Other genera including *Diplocaulus* and the captorhinids increased sharply in numbers, responding to environmental circumstances to which they were well suited. In both instances speciation occurred during the Clear Fork.

None of the large prey animals maintained their relative abundance, failing to adapt to the changing circumstances. Either they became extinct during or soon after the initiation of Clear Fork Deposition (*Ophiacodon*, *Edaphosaurus*) or continued in relatively small numbers (*Eryops*, *Diadectes*).

With the climatic changes the aestivators *Gnathorhiza* and *Lysorophus* became prominent members of the Clear Fork chronofauna. They assumed greater importance in the vertebrate food web, but apparently neither became a primary food source for *Dimetrodon*.

To the extent that the successful genera were close associates in the tight Clear Fork chronofauna, they may be considered to have followed a course of coevolution, in a loose sense of the term. It is not demonstrable, however, that their relationships did involve the evolutionary interaction between species which are implicit in the idea of coevolution as used in the paper (fig. 2). Rather the evidence suggests a coadaptational relationship.

The inferred intimate predator-prey relationship between *Dimetrodon* and *Diplocaulus* offers the best chance of detection of coevolutionary relationships. If there were such a relationship it would follow that congruent changes would occur as the result of reciprocal selective interactions. *Diplocaulus* did undergo detectable although slight morphological changes with an accompanying shift from still to running water habitats. This habitat shift may have posed behavioral problems for *Dimetrodon*, but the morphological changes alone certainly were not of the sort that would have affected the predator-prey relationship and elicited compensating morphological changes in *Dimetrodon*. It is clear that this did not in fact occur unless there were subtle, undetected changes in the postcranium, jaws and teeth of *Dimetrodon*.

The close associations of *Dimetrodon* and *Diplocaulus* continued without change after *Diplocaulus* had assumed a new way of life. Thus it must be assumed that *Dimetrodon* did adapt to the new circumstances presented by its prey and that both members of the pair system underwent coadaptation to the new circumstances. In the case of *Diplocaulus* the response was evolutionary, in the case of *Dimetrodon* it appears to have been strictly behavioral. Had the close association of the two genera in samples failed to continue to persist, the hypothesis of coadaptation would tend to be falsified.

Similar explanations may apply to the relationships of *Dimetrodon* to other presumed prey, but the criteria are somewhat less reliable. Both *Dimetrodon* and *Trimerorhachis*, for example, adjusted to changing environmental conditions without evident morphological changes. *Seymouria* and the captorhinids both underwent speciation. In the former case changes were primarily in size and accompanying allometric alterations. In the latter, size also increased, but extensive modifications in dentition indicate dietary changes, apparently in adaptation to increasing opportunities for terrestrial herbivores. The importance of these larger animals as a source of food to *Dimetrodon* increased relative to that of their Arroyo predecessors, but no functional morphological changes were required in the predator. A coadaptational relationship appears to have existed as both became increasingly adjusted to changes in climate, one group by evolutionary modifications and the other by behavioral and physiological changes which may or may not have had an evolutionary base. Once again, these congruent changes may represent in a loose sense coevolution, but there is no impelling evidence that this was the case.

SUMMARY AND CONCLUSIONS

The chronofauna of the Clear Fork, a tightly knit component of the Permo-Carboniferous chronofauna, consists of a closely interwoven series of organisms each of which plays one or more roles in the food chain. So portrayed the Clear Fork complex of vertebrates may suggest that the system persisted as an integral unit through coevolutionary relationships of its constituents. The system and its Wichita predecessors have been examined in the perspective of the relationships of the top predator to its principal prey. The analysis shows the major difficulties in reaching clear and conclusive conclusions concerning the evolutionary relationships of the system, even though the data are among the best available for this stage of development of terrestrial vertebrates.

A major change in the composition and structure of the complex took place with the onset of Clear Fork deposition, a change related to a strong shift toward drier climates. It presumably involved in particular a strong modification of the predatory habits of *Dimetrodon*. The shift produced the stable Clear Fork chronofauna.

During the existence of this chronofauna *Dimetrodon* and *Diplocaulus* were closely associated in predator-prey relationship. *Diplocaulus* underwent speciation accompanying a habitat change. Some of the terrestrial prey animals of the Clear Fork also underwent evolutionary changes, to some extent replacing ecologically their predecessors which had become extinct or greatly reduced. *Dimetrodon*, although surely making adjustments to the new climatic regime and changes in habitat of one of its principal prey animals, does not seem to have undergone any morphological change. Its adjustment to the climate and to the habitat shift of *Diplocaulus* is best considered in the context of coadaptational relationships between the two. That coevolution took place cannot be ruled out, but a hypothesis of coevolutionary changes cannot be falsified on the basis of data now available.

Other predator-prey relationships involve both genera in which there was no evident functional-morphological change and those in which there was evolutionary change. These suggest as well that the relationships were primarily coadaptational with the top predator and that it was coadaptation that maintained the integrity of the chronofauna until its demise, after steady loss of constituents, during the late phases of deposition of the Clear Fork beds.

ACKNOWLEDGMENTS

Dr. Peter P. Vaughn read early drafts of this manuscript and made valuable suggestions which have been incorporated in the text. Dr. Kathryn Bolles prepared the illustrations. Financial support was given by a grant from the National Science Foundation, DEB 80 00922. For this aid I express my gratitude and thanks.

LITERATURE CITED

BARGHUSEN, H.R. 1973. The adductor jaw musculature of *Dimetrodon* (Reptilia: Pelycosauria). Journal of Paleontology, 47:823-834.

CASE, E.C. 1915. The Permo-Carboniferous red beds of North America and their vertebrate fauna. Carnegie Institution of Washington Publication no. 207:1-176.

CASE, E.C. 1919. The environment of vertebrate life in the late Paleozoic of North America: a paleogeographic study. Carnegie Institution of Washington Publication no. 283:1-273.

DOUTHITT, H. 1917. The structure and relationships of *Diplocaulus*. Contributions from Walker Museum, 2:1-41.

OLSON, E.C. 1951. *Diplocaulus*, a study in growth and variation. Fieldiana: Geology, 11:57-154.

OLSON, E.C. 1952a. The evolution of a Permian vertebrate chronofauna. Evolution, 6:181-196.

OLSON, E.C. 1952b. Fauna of the Vale and Choza: 6. *Diplocaulus*. Fieldiana: Geology, 10:147-156.

OLSON, E.C. 1955. Fauna of the Vale and Choza: 10. *Trimerorhachis*, including a revision of pre-Vale species. Fieldiana: Geology, 10:225-312.

OLSON, E.C. 1958. Fauna of the Vale and Choza: 14. Summary, review and integration of the geology and faunas. Fieldiana: Geology, 10:397-448.

OLSON, E.C. 1975. Permian lake faunas: a study in community evolution. Journal of Paleontological Society of India, 20:146-162.

OLSON, E.C., and J. MEAD. 1982. The Vale Formation (Lower Permian): its vertebrates and paleoecology. Texas Memorial Museum, Bulletin 29:1-46.

OLSON, E.C., and P.P. VAUGHN. 1970. The changes of terrestrial vertebrates and climates during the Permian of North America. Forma et Functio, 3:113-138.

ROMER, A.S. 1928. Vertebrate faunal horizons in the Texas Permo-Carboniferous red beds. University of Texas Bulletin 2801:67-108.

ROMER, A.S., and L.I. PRICE. 1940. Review of the Pelycosauria. Geological Society of America Special Paper 28:1-538.

VAUGHN, P.P. 1969a. Lower Permian vertebrates of the four corners and midcontinent as indices of climatic differences. Proceedings of the North American Paleontological Convention, part D:338-408.

VAUGHN, P.P. 1969b. Early Permian vertebrates from southern New Mexico and their paleozoogeographic significance. Contributions in Science, Los Angeles County Museum, no. 166:1-22.

WILLIAMS, M.E. 1972. The origin of "spiral coprolites." Paleontological Contributions, University of Kansas, paper 59:1-19.

WILLISTON, S.W. 1911. American Permian vertebrates. Chicago: University of Chicago Press:1-145.

MODELS OF COEVOLUTION: THEIR USE AND ABUSE

Montgomery Slatkin

Department of Zoology
University of Washington
Seattle, Washington

The use of mathematical models in the study of coevolution is illustrated in the case of ecological character displacement. The biological background to the problem and some of the theoretical studies are reviewed. Contrary to the prevailing biological intuition, competition among sympatric species does not inevitably lead to character displacement. Instead character displacement results from the imbalance of forces, one being the force of competition between species that tends to select for differences and the other being the force of competition within each species that tends to increase the extent of phenotypic variation and lead to character convergence. Mathematical models can show how these forces balance each other under different biological assumptions but they cannot settle the issue of what actually has happened in natural systems. To imply that models can resolve this and similar issues is to confuse mathematical results with biological intuition.

INTRODUCTION

Coevolution, defined broadly as the evolution of two or more species as a consequence of their interactions with one another, has tremendous intuitive appeal. The struggle for existence that Darwin described is largely a struggle against members of other species. If natural selection has been the predominant agent of evolutionary change, then coevolution must be commonplace. One task of evolutionary biologists is the description of how and at what rates coevolution occurs. Mathematical models of coevolution serve very well the purpose of charting the course of coevolution and largely confirming the intuition that coevolution can occur easily and rapidly. While mathematical models are useful in formalizing intuition and working out the consequences of different biological assumptions, they also can be abused when more is attributed to them than has gone into their making. Models can describe what may happen but not what has happened.

I will discuss here one type of coevolution, that of ecological character displacement, which has been widely discussed and modeled. I will try to show how models have been used to lead to a better understanding in this area and how they can be abused to lead to unjustified conclusions.

ECOLOGICAL CHARACTER DISPLACEMENT

The term "character displacement" was introduced by Brown and Wilson (1956) to describe the situation in which two species exhibit morphological differences in sympatry but not in allopatry. The idea is that populations in sympatry and allopatry represent a natural experiment, allowing the conclusion that differences can be attributed to the presence or absence of the other species. Brown and Wilson's original discussion was in terms of a morphological trait like body size or bill size in birds, with the sympatric populations being more different than the allopatric ones, as shown in Figure 1. Grant (1972) has argued for a more general definition of the term for cases in which sympatric populations may be more similar than allopatric ones or differ in some other way.

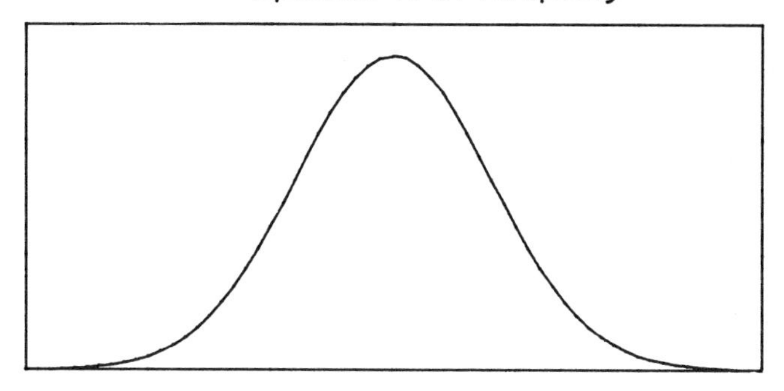

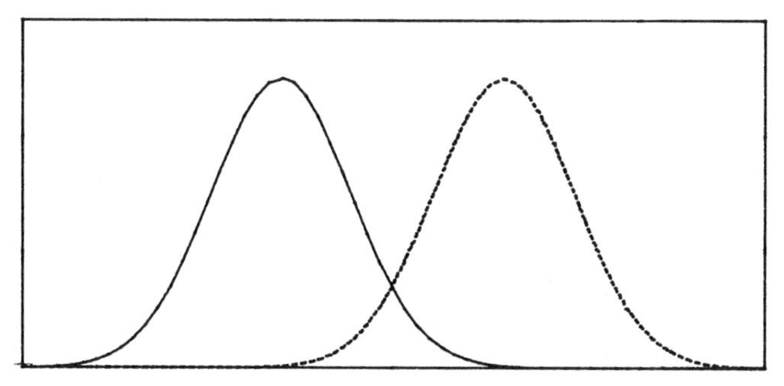

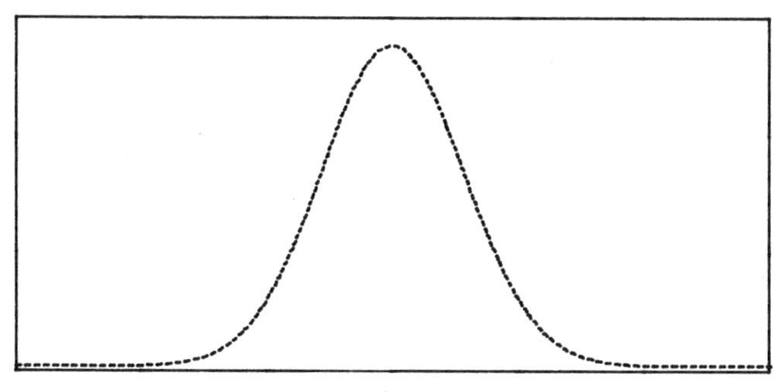

Figure 1. An illustration of the basic idea of character displacement. If the populations of the two species are different in allopatry and sympatry, the difference is attributed to the presence of the other species in the region of sympatry.

This more general definition of character displacement has been partly adopted and the term is often used to describe any evolutionary changes caused by the presence of other species (Lawlor and Maynard Smith, 1976; Dunham, Smith and Taylor, 1979). Here, I will use it in the older sense to mean divergence.

Brown and Wilson (1956) distinguished two mechanisms for causing character displacement: competition and assortative mating. Competition led to "ecological character displacement" and assortative mating to "reproductive character displacement." Ecological character displacement quickly became a textbook illustration of the effectiveness of competition in natural populations. I have used it myself for that purpose many times. The term originally was used for only two species but it could be and was applied to cases with more. The classic study of the ground living Darwin's finches (*Geospiza*) studied by Lack (1947) became one of the common illustrations of ecological character displacement.

Ecological character displacement was originally an evolutionary phenomenon, since it is an ecological mechanism that accounts for particular geographic patterns of morphological variation. Hutchinson (1959) reversed the argument and used the evolutionary process to account for the ecological phenomenon that, within sympatric groups of related species, there tended to be consistent differences in size. In particular, Hutchinson (1959) argued that a ratio of sizes of about 1.3 is a minimum necessary to permit cooccurrence of competing species. In Hutchinson's view, competition leads to character displacement and the resulting differences among species permit their coexistence. The two processes of morphological diversification through character displacement and the exclusion of species that compete too intensely formed the major part of the answer to Hutchinson's question: "Why are there so many kinds of animals?" The idea that competition limited the number of species in natural communities also became part of the textbook orthodoxy.

The possibility of ecological character displacement was sufficiently agreeable to the collective intuition of ecologists that it was accepted long before it was modeled. It did and probably still does seem sufficiently obvious that, if two or more species are competing, natural selection will tend to reduce the extent of competition among them. There was also no shortage of examples of ecological character displacement, found not only among extant species but among fossil species as well. The theory, albeit non-mathematical, was confirmed by observations and the interpretation of the observations was buttressed by the theory. It was an ecological success story that fueled a generation of Ph.D. candidates.

More recently, the ubiquity of ecological character displacement has been challenged. The first major attack was by Grant (1972, 1975) who questioned the premise of Brown and Wilson's logic, namely that differences between allopatric and sympatric populations can be attributed entirely to the presence or absence of the other species. There can be many factors that vary in space, and populations of the two species could be responding to those factors, not to each other. One possibility described by Grant (1975) is illustrated in Figure 2. Both species show clines in size. The differences in body size in the region of sympatry may not be due to the presence of the other species at all but to some other environmental factor that is responsible for the clines. It is possible, of course, that the factor is competition from the other species so a pattern of this type does not

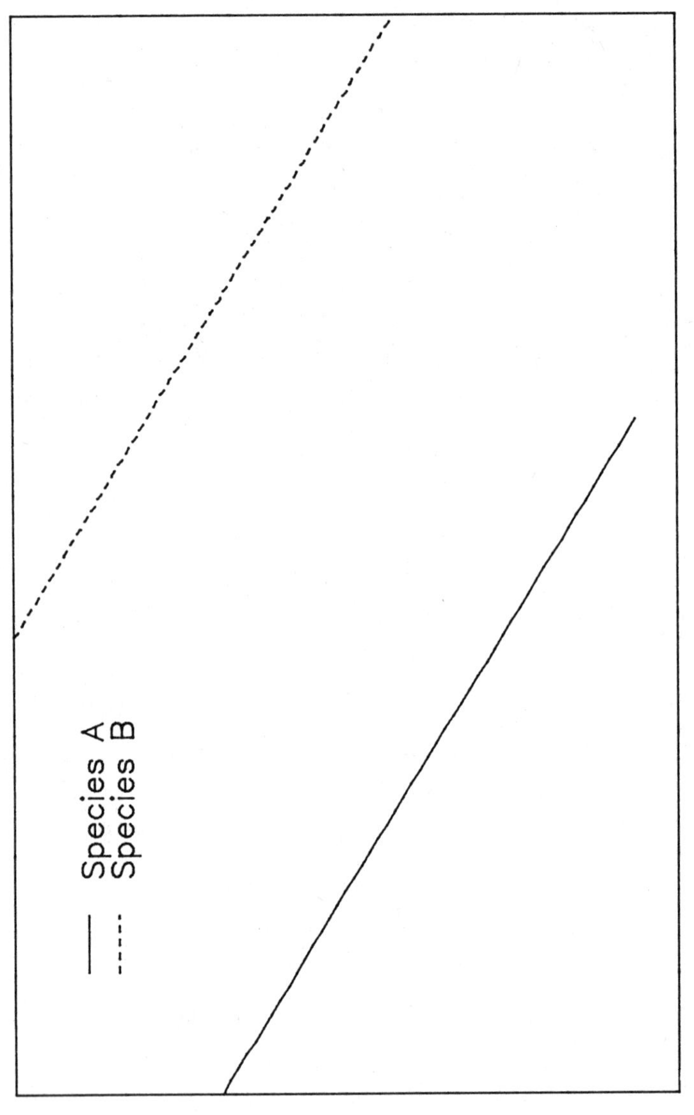

Figure 2. An illustration of how a clinal pattern in each of two species could produce the pattern that would suggest that character displacement had occurred. The species would differ in allopatry and sympatry but the differences could be due to the geographic variation in an environmental factor that produces the cline in each species (after Figure 4 in Grant, 1972).

necessarily refute the mechanism of competition, but it alone does not confirm it either. Other possible explanations would have to be excluded, and that would require some other kind of study.

Grant's objection also agrees with intuition. If the habitats of the sympatric and allopatric species are sufficiently similar that the situation can be viewed as a natural experiment, then the question arises of why the species are ever allopatric. If they can evolve in response to each other in some areas and coexist as a result, why have they not done so everywhere. This question seems particularly relevant for the examples discussed by Brown and Wilson (1956) of birds that would appear to be able to reach the habitats where they are not now found. The fishes analyzed by Dunham et al. (1979) are in different drainage systems so this objection is probably less important. Their analysis shows one way to tease out the variation in a trait caused by competitors and by geographic variation in the environment.

The second kind of attack on ecological character displacement was not directly on the problem discussed by Brown and Wilson (1956) but on Hutchinson's (1959) generalization. Simberloff, Strong and their co-workers have constructed "null models" to predict some aspects of community composition under the assumption that competition is not occurring (e.g., Simberloff and Boecklen, 1981; Strong, Szyska and Simberloff, 1979). The null models provide a way to ask the same question as was asked by Grant, namely can the pattern in the data be accounted for without invoking competition?

A particularly good example of the use of null models is by Simberloff and Boecklen (1981). They analyze the same kind of data as was used by Hutchinson (1959) to show that competition was effective. Their null model is that the logarithms of the average sizes of species in a community are uniformly distributed. With this null model, they generate the distribution of ratios of sizes and the distribution of minimum size ratios for different numbers of species. They then examine numerous data sets to determine whether they differ significantly from the pattern generated by the null model. They found that most data showed no significant differences from the expectation under the null model and several showed that the sizes tended to be more similar than expected under the null model. As discussed by Simberloff and Boecklen (1981), many people had "confirmed" Hutchinson's (1959) prediction that there tended to be relatively constant ratios of body sizes among species supposed to be in competition despite the fact that their data showed that the distribution of the body sizes could be completely random. The point made by Simberloff and Boecklen (1981) is the same as made by Grant (1972). Observations may be consistent with the biological assumption that competition is responsible for a pattern, but that does not prove that competition has been effective. Other mechanisms and possibly no mechanism at all may account for the same observations. These studies and others call into question the ubiquity of ecological character displacement as a biological phenomenon.

The statistical model of the type developed by Simberloff and Boecklen (1981) can be used to establish whether there is a significant pattern in the distribution of a character in a group of sympatric species. If we assume that for a particular group there is a pattern, then there are two different but not entirely separate classes of events that are relevant to account for that pattern. The first class is of the evolutionary events that led to the differences between the species in the character of interest. The second is of the ecological events that led to the cooccurrence of the species and that permit their continued coexistence. I discuss evolutionary models first and then the ecological models. While they are different kinds of models and are directed to different questions, both types must be examined to understand the theoretical basis for ecological character displacement.

EVOLUTIONARY MODELS

In this and the next section, I discuss mechanistic models, which are models of particular processes based on biological assumptions about how those processes work. A mechanistic model is developed in several steps. First, the biological assumptions about the process are needed. The biological assumptions must then be refined sufficiently that they can be translated into mathematical assumptions that have the same meaning. Usually the most difficult stage is proceeding from a biological discussion of a process, in which only some of the assumptions are made explicit, to the mathematical assumptions.

The next step is working out the consequences of the mathematical assumptions. Many biologists regard this as the most difficult step and indeed the only step. That is an unfortunate view of mathematical modeling but one that is shared by some mathematicians. It is not a conceptually difficult step because mathematical assumptions, if they are clearly and unambiguously stated, lead inevitably to certain conclusions. It may require extensive labor to obtain results from a mathematical model, with some models yielding their conclusions only to very subtle and elegant methods of the pure mathematician or to the inelegant brute force of computer simulations. The difficulties in reaching the conclusions should not be confused with the certainty of reaching them.

The last step in the process also can be difficult. The predictions of the model, which are in mathematical terms, must be translated back to the biological system being modeled. It is only then that the biological value of the model can be assessed.

Biological Assumptions

It is relatively easy to summarize in general terms the biological assumptions underlying ecological character displacement. In each species, there is assumed to be a heritable character which in some way determines which food resources individuals can efficiently use. Common examples are characters like body size or bill size that restrict the range of prey sizes an individual can consume. Food resources are assumed to be limited so the extent of competition between two individuals is partly determined by the difference between them in the character. Individuals that are more similar compete more intensely.

To state these assumptions precisely so they can be put in mathematical form calls for making explicit assumptions that are usually not a part of a biological discussion. The character of interest is assumed to be heritable, but to construct a model some further assumptions about its genetic basis are needed. The character could be controlled by a single genetic locus; it could be controlled by several loci whose effects are additive; or it could be inherited in some other way. Most discussions of ecological character displacement are of characters that are approximately normally distributed in each species, although the full distribution is usually not discussed. The assumptions about the genetic basis of the character determine how the mean and variance of the character can change. The most general possible model one that assumes the mean and variance can both change. That is what results if we assume that the character is controlled by a large number of loci with additive effects, the usual quantitative genetics formulation. On the other hand, in some species the variance may not be able to change at all or can change only in restricted ways. If that is so, then some other assumptions about the character are more appropriate. We see that the general notion of a heritable character allows for several possible specific assumptions.

There also are several ways to interpret the assumption that the difference between two individuals determines the extent of competition between them. The intensity of competition between two individuals in the same species could be more or less intense than between two comparable individuals of different species. Individuals of different species could be competing for some resources, but each species could have exclusive access to some others. Or they could both depend completely on the same resources. These possibilities all embody the idea that the two species are competing. A mathematical model and especially the predictions of a particular model cannot answer the question of which assumptions are more appropriate.

Mathematical Assumptions

I will summarize the mathematical assumptions that represent the biological assumptions just described. While I will concentrate on the assumptions I used in a model of ecological character displacement (Slatkin, 1980), I will relate them to those made in other models. I emphasize again that there is no single "right" set of assumptions to be made. Different assumptions simply model different situations.

In each species, we assume that the character can be measured on a single scale, which is chosen so the character is normally distributed. If there are two species, species 1 and species 2, let the average of the character in species i be $\bar{z}_i$ and let its variance be V_i. Assumptions about the genetic basis of the character lead to predictions about the mean and variance among the offspring of a particular group of parents. In very general genetic models, the mean and variance in one generation are not sufficient to predict those values for the next generation. The frequencies of alleles at different loci and the linkage disequilibria among the loci would be needed. However, if simple assumptions are made, then the mean and variance in one generation can be used to predict the values in the next. One model that has this property is a model that assumes the character is controlled by a single genetic locus with only two distinct alleles. The mean (or the variance) of the character is completely determined by the allele frequency, so if the mean and variance are known in one generation, they can be predicted for the next. In this model the mean and variance cannot change independently. They are both determined by the value of the frequency of one of the alleles (Slatkin, 1979). Instead, if there are many possible alleles at the locus, we can arrive at a model similar to that developed by Kimura (1965), in which there are a potentially infinite number of distinct alleles with additive effects. For this model, the mean and variance change according to the equations

$$\bar{z}' = (1 - h^2)\bar{z} + h^2(\bar{z} - \bar{z}_s)$$
$$V' = (1 - h^4/2)V + (h^4/2)V_s \qquad (1)$$

(assuming that generation are discrete and non-overlapping) where the primes indicate the values in the next generation and the subscript s indicates the values after selection but before random mating. Lande (1976) has generalized Kimura's approach to more than one genetic locus. Lande showed that the covariance between the frequencies at the different loci, which are equivalent to the linkage disequilibria, also can affect the changes in the mean and variance. However, if the covariances are small, the effect is also small. My approach was to assume a single locus model, Kimura's model, and then later consider the possible effects of linkage disequilibrium and other genetic factors. This approach was used, not because it was the most realistic possible, but because it was simple enough to allow a fairly general analysis and still allow for a discussion of other genetic factors.

Another possible approach to modeling the inheritance of a quantitative character is taken by Roughgarden (1976) in his model of character displacement. He assumed that the variance of the character did not change and that only the mean was under genetic control. This is not the same as a one locus model, even if the mean is assumed to be controlled by a single genetic locus. In the one locus model, which was used by Fenchel and Christiansen (1977) and Crozier (1974), both the mean and variance can evolve although not independently. Roughgarden's approach is interesting because it is based not on *a priori* assumptions about the genetic basis of the character but on the empirical claim that variances of most characters do not change by much. Roughgarden's generalization is not always true. For example in Darwin's finches in the Galapagos Islands, Grant and Price (1981) show that variances of bill size in populations of a species on different islands differ, but the variance in body size of *Anolis* lizards on different islands in the West Indies does not change by much (Roughgarden, 1974).

Yablokov (1974) summarizes available data on mammals to show that the coefficient of variation in phenotypic characters (the square root of the variance divided by the mean, a dimensionless way to measure the extent of variation in characters with different means) is similar for the same characters in different species but different for different characters in the same species. For example, the coefficient of variation in postcranial skeletal characters in mammals is typically about 10% while the coefficient of variation for the length of the digestive system is usually 15% or larger (Yablokov, 1974). If this is true generally and the variance of most phenotypic characters does not evolve or evolves only very little, then Roughgarden's (1976) assumption is valid. The validity of this assumption is not established by the model. Even if the predictions of a model developed making this assumption agree with observations or with intuition, that does not validate the assumption.

There are two ways to proceed in developing a model of competition. The first is to make some assumptions directly about the intensity of competition between different individuals in terms of the effect of one individual on the fitness of another. The second is to make more detailed assumptions about how individuals obtain the resource, how efficiently they convert the resource into offspring and survival, and how they interact with one another to reduce the amount of the resource each can gather. This approach leads to the development of an "inner" model of resource utilization that is itself based on a number of assumptions and that leads to predictions about the way individuals compete. There is sometimes no difference in the actual form of the competition based on the character. A model of competition that is baldly assumed --- the first approach --- could also be derived from some assumptions in an inner model -- the second approach. That a model of competition can follow from mechanistic assumptions about the behavior and ecology of the species being considered lends some confidence to that model, but we should not lose sight of the fact that it is still a model. Without independent verification of the assumptions of an inner model, either approach to modeling is simply an expression in mathematical terms of the biological assumptions about competition.

In modeling ecological character displacement, there is fortunately no great difference between a simple model of competition that is assumed and a model derived from an inner model. One way to develop a model and the way that most workers have followed is to extend the ideas in the Lotka-Volterra competition equations (Roughgarden, 1979). We can recall that in modeling competition between two species, a common approach in theoretical ecology is to assume that there is some conversion factor that relates the effect of individuals of one species on those of another to the effect of members of the other species on themselves. The "competition coefficient", which is often denoted by α, is the measure of the ratio of these effects. If $\alpha>1$, one individual of the other species has a greater effect on an individual of the first species than does a member of the first species on a conspecific. If $\alpha<1$, the opposite is true and an individual of the other species has a smaller effect. If $\alpha = 1$, members of the two species are equivalent in their effects. This way to model competition between two species, each of whose members are regarded as being the same in the competitive effects, does not assume anything about the mechanism of competition. Individuals may be depriving each other of resources by consuming them, by fouling them, or by preventing others from gaining access to them. Or competition could be due to direct aggressive interactions between individuals. MacArthur (1968) showed that it is possible to construct an inner model, in which members of two species compete for two limiting resources, that leads exactly to the Lotka-Volterra competition equations under certain conditions.

Following Roughgarden (1972) we can construct a model of competition based on the character of interest by generalizing this approach. If we consider an individual of type z in species i and an individual of type z' in species j, then we can define $\alpha_{ij}(z,z')$ to be the ratio of the reduction in fitness of the second individual by the first to the reduction in fitness of the first individual by one of the same type. From this definition $\alpha_{ii}(z,z) = 1$. This assumption about competition allows for many more possibilities than are needed for the discussion of ecological character displacement.

The biological assumptions underlying ecological character displacement imply that the intensity of competition between two individuals decreases with increasing difference in the character. That suggests that we can restrict ourselves to the consideration of a competition function of the form $\alpha_{ij}(z,z')$, with only the difference between the characters of the two individuals determining the extent of competition between them. Moreover, we can assume that this function decreases as the absolute value of the argument increases. We can simplify further and assume that the competition depends on the difference between the values of the character and on whether the species have the same or different identities. That is, we assume that $\alpha_{ij}(z-z') = \alpha(z-z')$ if $i = j$ and $\alpha_{ij}(z-z') = C\,\alpha(z-z')$ if $i \neq j$. This way of expressing our assumption separations competition into two parts. The function $\alpha(z-z')$ expresses how competition depends on the relative values of the characters and the constant C determines whether competition is more or less intense within or between species. If $C>1$, competition is more intense between species than within, with the reverse being true if competition $C<1$. If $C = 1$, then competition depends only on the character and not on the specific identity.

The biological assumptions about competition do not indicate which range of values of C is the most important. Many workers agree with Darwin (1859, Chapter 3) that competition is more intense between members of the same species. That would imply $C<1$. However, it is also possible to argue that there could be social or other facilitation between members of the same species that would not occur between members of different species. That would suggest that $C>1$. It is worth considering different values to find if some will lead to character displacement while other will not. Most models of character displacement assume that $C = 1$. Roughgarden (1979) discusses how this assumption of competition can be related to an inner model similar to that of MacArthur (1968).

The last part of a mechanistic model needed is an assumption about the abundance of resources the two species are competing for. Once again using the Lotka-Volterra competition model as a starting point, we recall that the abundance of resources is expressed as the carrying capacity, usually denoted by k. This is the number of individuals the environment could support if there were no competition from the other species. Generalizing this to the case in which individuals can be distinguished by the character, we can define $k_i(z)$ to be the number of individuals in species i

that could be maintained if they could reproduce themselves. While the genetic assumptions about the character imply that individuals of type z will not produce individuals only of that type, we can imagine they could, just for the purpose of describing the abundance of resources.

To summarize the model of competition, the two species are competing for resources whose abundances are measured by the two functions $k_1(z)$ and $k_2(z)$, with the extent of competition determined by the function $\alpha(z-z')$. To make specific predictions about the genetic evolution resulting in each species, we have to assume forms for these functions. To test the generality of the results obtained, we have to consider a variety of functional forms. Particularly simple forms that have been used by several people are based on the normal distribution. Let the carrying capacities be given by

$$k_i(z) = K_i \; exp[-(z-x_i)^2/2\sigma_k^2]/\sqrt{2\pi\sigma_k^2}. \quad (2)$$

Equation (2) has three parameters, K_i, which indicates the total abundance of resources, x_i, which indicates the value of the character for which there are the most resources, and σ_k, which indicates the diversity of resources available. If σ_k is small, then there is a narrow range of resources available. if $x_1 \neq x_2$, then the two species are competing for different resources or at least are using the same resources in different ways. The carrying capacity function is graphed in the top part of Figure 3. This functional form of $k_i(z)$ assumes that there are equal resources available for individuals with small and large values of z. If that is not true and there are, for example, many more resources available for small rather larger values of z, then the non-symmetric form shown in the bottom part of Figure 3 would be more appropriate.

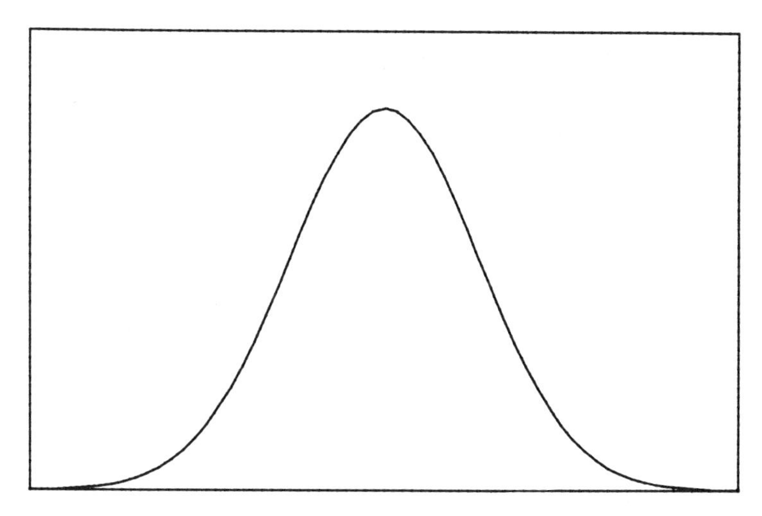

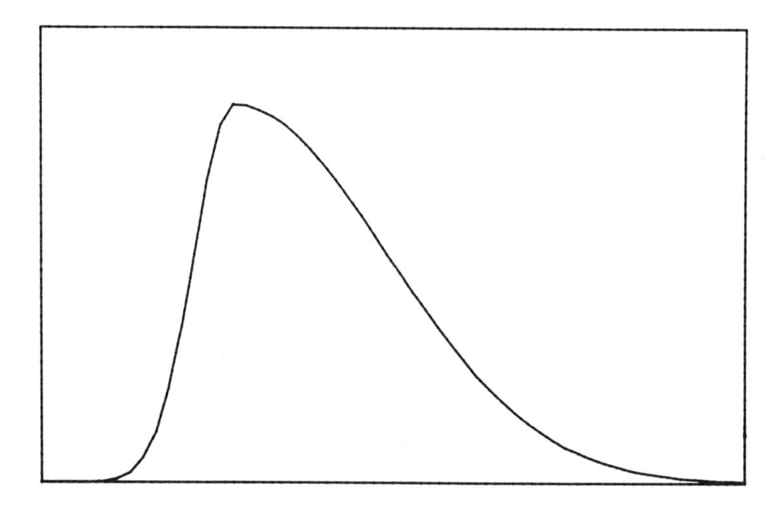

Figure 3. Examples of a symmetric and assymetric resource distribution function, as described in the text.

A simple functional form for the competition function is

$$\alpha(z-z') = exp[-(z-z')^2/2\sigma_\alpha^2] \tag{3}$$

This has only a single parameter, σ_α, which indicates how different two individuals must be for there to be little or no competition between them. This functional form assumes that competition is symmetric as shown in the top part of Figure 4. If individuals with larger values of z compete more intensely with those with smaller values than the reverse, then the asymmetric form shown in the bottom part of Figure 4 is more appropriate.

Predictions of the Evolutionary Model

At this point you probably have a rather uneasy feeling produced by two opposing dissatisfactions. On one hand the model I have described obviously is extremely complex, bristling with enough definitions and parameters that you might have forgotten most of them already. On the other hand, I have had to make so many assumptions that there appears to be little left of the biological problem. These dissatisfactions both are well founded, and what you suspect about this model is true. It is both complex and based on numerous simplifications. My defense is that the complexity of the model reflects the underlying complexity of the biological problem. And the simplifying assumptions are no different than the assumptions made in intuitive discussions of the problem; it is just that a mathematical model must make all those assumptions explicit. An intuitive discussion has the luxury of being able to make implicit assumptions; a mathematical model does not.

The model was developed to answer the question: does competition for a limiting resource lead to character

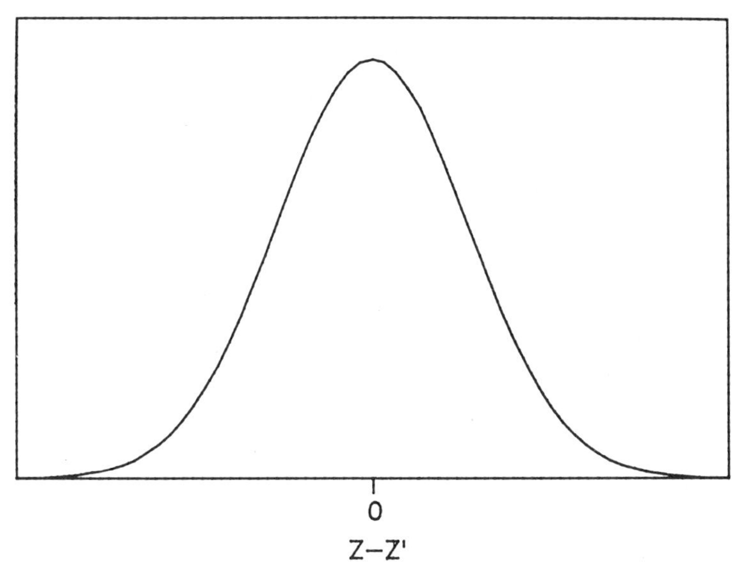

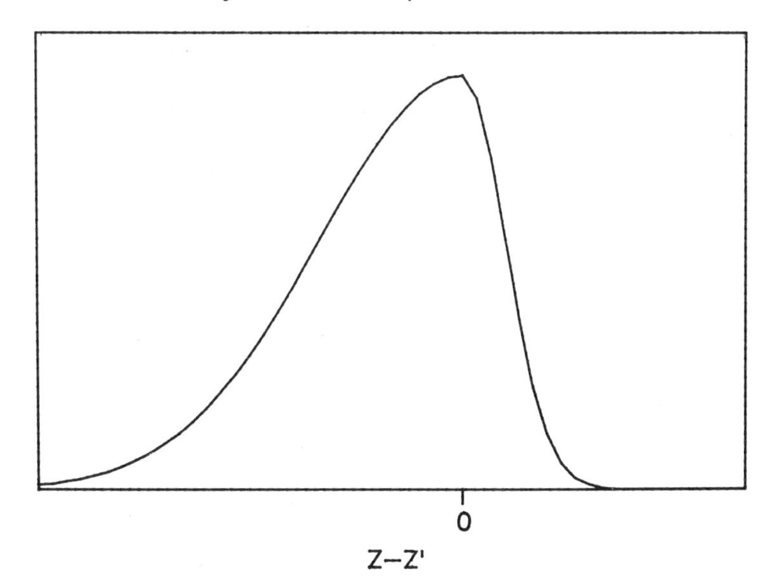

Figure 4. Examples of a symmetric and asymmetric competition function, $\alpha(z-z')$*, as described in the text.*

displacement? To summarize a rather large number of computer simulations of this model, the answer can be stated with great confidence: sometimes yes, sometimes no. It should not be too surprising that such a complex model does not always produce the same outcome. What was more surprising, to me at least, was that some of the simplest assumptions, including those that other people had also made, did not inevitably lead to significant character displacement.

To start with the simplest case, I assumed that $k_1(z) = k_2(z)$ (i.e., the two species are competing for exactly the same resources) and $C = 1$ (i.e., competition depends only on the difference in the character and not on specific identity). These or very similar assumptions were made by Bulmer (1974), Fenchel and Christiansen (1977) and Roughgarden (1976). who all concluded that significant character displacement could occur. By significant, they meant and I mean that the difference between the means of the characters in the two species (sometimes denoted by $d = \bar{z}_1 - \bar{z}_2$) is about 2 or 3 times the standard deviation of the character in both species, which is sometimes denoted by w (V_1 or V_2 assuming they are roughly equal). A value of 2 or 3 for the d/w ratio is typical of the observed difference among species like the Darwin's finches, which are presumed to have undergone ecological character displacement. For a model to explain the observations, it has also to produce a d/w ratio of 2 or 3.

In my simulations, I found that I did not get a value of d/w much greater than 1 in most cases. And for some initial conditions, I found that these assumptions led to character convergence, not divergence. When I say that the same assumptions as used by others, led to different conclusions, I mean only that the assumptions about the competition and resource abundance were the same. I used a more general model of inheritance, although not the most general one possible, and that turned out to be the source of the differences between my results and those of other models. My model of inheritance allowed both the mean and variance to change and the differences between the models were due to changes in the variance. While intuition suggests that evolution should always act to reduce the intensity of competition between species, that intuition does not account for the competition among members of the same species. While the mean value of the character does tend to respond mostly to competition between species, the variance responds to competition within species as well. The effect of within-species competition is to change the variance to the point at which there is no more selection on the mean value (Slatkin, 1980). I verified that this was the source of the difference between my results and others by adding a constraint on the variances. I assumed that the variance could not increase beyond a certain limit. If that limit were small enough, then it was easy to obtain values of d/w of 2 or 3 or even larger. At this point, Roughgarden's (1976) assumption that the variance within each species remained fixed becomes important. If it is fixed at a small enough value, substantial character divergence occurs. But the divergence is due not only to the competition between species but also to the constraint on the variance. The model discussed here predicts character displacement only under circumstances more restricted than biological intuition would suggest.

This model has other features that I summarize but not elaborate on. I have already shown that the simplest assumptions do not necessarily lead to character displacement, and it is worth exploring the conditions under which it will occur.

If the distribution of resources is not symmetric, as shown in the bottom part of Figure 3, then I found that significant displacement would occur under many parameter values (Slatkin, 1980). There were two differences of interest from the results for the symmetric case. First, the abundances of the two species are very different, with the species at the value of z for which the most resources are available being much more common than the species at the tail of the curve. Second, assuming that the variances are allowed to evolve, the variance of the more abundant species is smaller than that of the rarer species. Unfortunately, no data exist on the relative abundances of species that are presumed to have undergone character displacement.

If the distribution of the resources are symmetric but with slightly different values of the maximal abundance, x_i, then I found that a substantial divergence would occur whether or not the variances were constrained, as long as they were not constrained to too large a value. The difference between the equilibrium mean values was substantially larger than the difference between x_1 and x_2.

If there are slight differences between the intensities of competition with and between species (i.e., $C \neq 1$), and if the variances are allowed to evolve, then I found the surprising result that the model was structurally unstable (Slatkin, 1980). If $C>1$, indicating that between-species competition is more intense than competition within-species, then one of the two species would go extinct. If $C<1$, then, there is character convergence, with both species evolving to the same equilibrium mean value.

If the competition function $\alpha(z-z')$ is not symmetric, as shown in the bottom part of Figure 4, then the results also depend on the assumptions about the variances but in a different way. If the variances are not constrained, then one species can drive the other to extinction, even when $C = 1$ (and for $C>1$ as well). If, however, the variances are constrained, the two species will coexist. There will be significant displacement if they are constrained to be small values but convergence if they are constrained to large values. The results are summarized in Figure 5. Roughgarden et al. (1983) use a different model of asymmetric competition and reach a different conclusion still. They find that if the variances are constrained (as Roughgarden (1976) argues they should be for biological reasons), then one species drives the other to extinction for some parameter values.

To summarize, there are several questions we can ask of this model of ecological character displacement. Does this model tell us whether character displacement will commonly occur? Well - no, it does not. Character displacement can occur under some conditions and not under others. The model cannot determine how common the right conditions are. Does the model tell us anything about ecological character displacement (a much more modest goal, indeed)? Well - yes, it does. It shows that ecological character displacement is not the inevitable outcome of competition between species. Instead, it depends on the balance between two forces, the force of competition between species tending to displace the mean values of the character and the force from the edges of the resource distribution tending to drive the mean values together. The relative strengths of the forces depend on the assumptions of the model and not in an obvious way. If the variances are constrained to very small values, then the balance is tipped in favor of between-species competition, because pressure from the edge of the resource spectrum cannot be felt. The result is divergence. If the variances

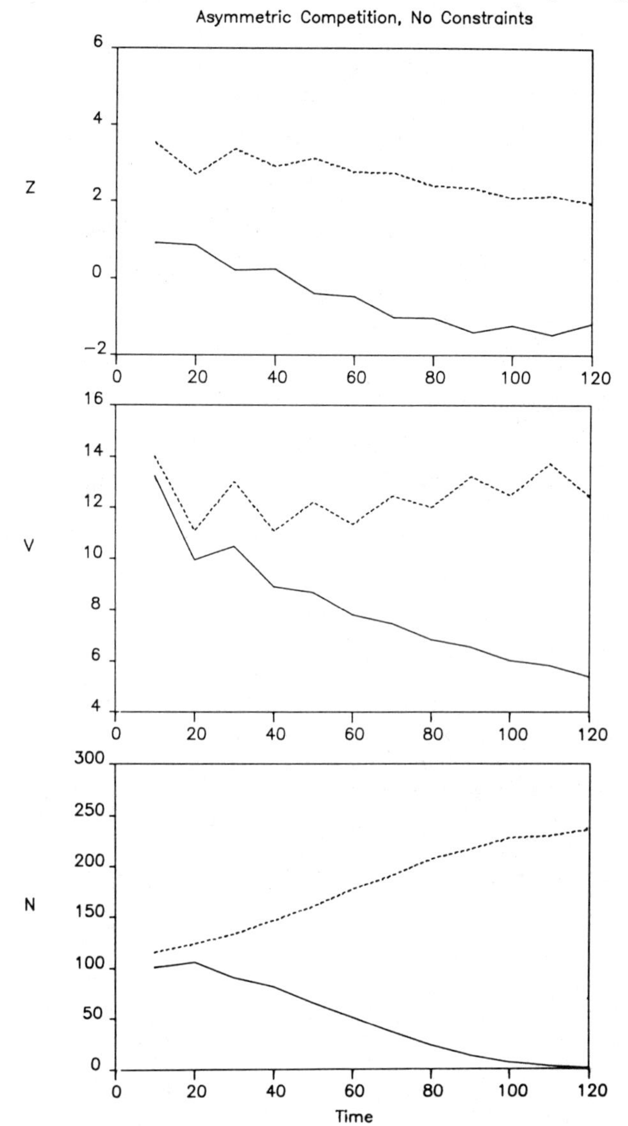

Figure 5. Graphical results for the case of an asymmetric competition function and symmetric resource distribution. When the variances are constrained, both species persist. The extent of displacement depends on how large the variances are. If the variances are not constrained, one or the other species can go extinct, although there are some initial conditions for which both species persist.

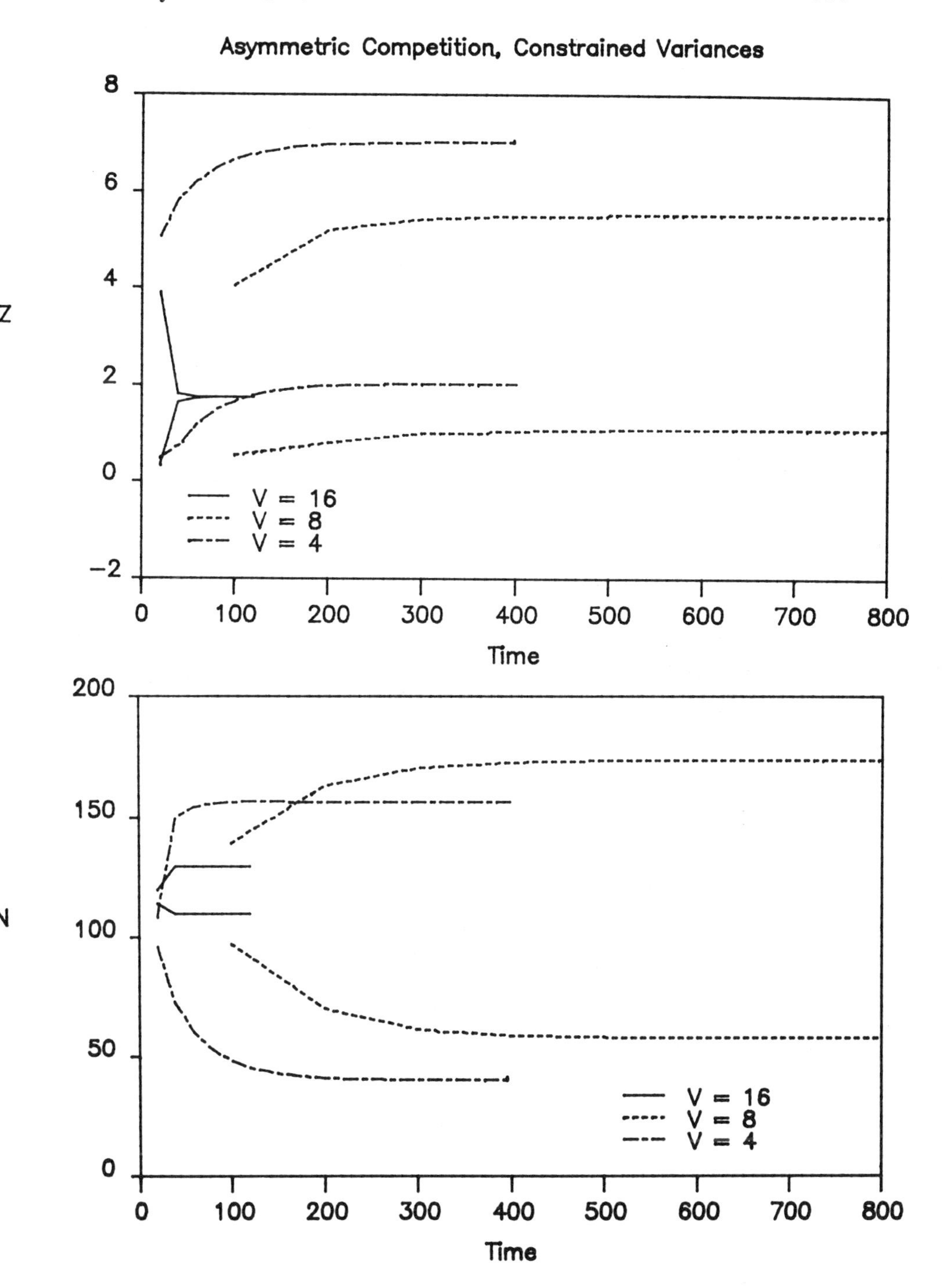
Asymmetric Competition, Constrained Variances
Z
8
6
4
2
0
−2
0
100
200
300
400
500
600
700
800
Time
V = 16
V = 8
V = 4
N
200
150
100
50
0
0
100
200
300
400
500
600
700
800
Time
V = 16
V = 8
V = 4

are constrained to large values, the balance is tipped the other way and the result is convergence. With an asymmetric resource distribution or with slightly different resource spectra for the two species, between-species competition has a slight advantage because each species can, to some extent, escape the effects of the other. The result is, again, divergence.

Demographic models of character displacement

In addition to the evolutionary models described in the preceding section, there are a variety of demographic models --models in which only the abundances of the competing species are allowed to change--that also are discussed in the context of character displacement. Demographic models cannot of course explain how differences between species evolved, but they investigate the outcome of the purely ecological interactions.

Demographic models come in two varieties that model two ecological processes. The first process is invasion. If a community is made up of some number of species with known phenotypes and those species have been together long enough for them to reach the equilibrium abundances, can another species of a known type invade the community? This question was first asked by MacArthur and Levins (1967). The answer is that a species can invade a community made up of two species if the mean values of the characters in those two species are sufficiently far apart (d/w greater than 2 or so). These results agree with the intuitive notion that competition will prevent too many similar species from being present in the same community. MacArthur and Levins (1967), in fact, developed their model partly to explain the general pattern found by Hutchinson (1959).

The other process is extinction. If there are several species in a community, will they tend to persist together? The answer from the deterministic ecological models--the same models that have been used so far--is yes. While competition can prevent the invasion of species that are too similar, it will not cause any species to go extinct, regardless of how similar they are. This clearly did not agree with Hutchinson's (1959) generalization about community structure.

May and MacArthur (1972) provided a theoretical basis for Hutchinson's (1959) generalization by showing that if there were assumed to be some environmental noise--random fluctuations in factors affecting all of the species--then extinction would result if the species were too similar. The May and MacArthur result agreed with that of MacArthur and Levins (1967) and with the prevailing intuition about the evolutionary process of character displacement.

We have already seen that the data do not show the pattern of character displacement or limiting similarity nearly as clearly as was once supposed and that the intuition about the evolutionary process is not nearly as trustworthy as we would have liked. It seems only fair then that May and MacArthur's (1972) conclusion was also not as general as was first claimed. Turelli (1977, 1978), in a dauntingly complex mathematical analysis of the problem posed by May and MacArthur showed that environmental fluctuations would not tend to cause the extinction of competing species under most conditions. If species do find themselves together in more or less equilibrium abundances, they will tend to persist together. Competition for a limited resource is simply not sufficiently intense to lead to extinction because, by assumption, each species is competing more intensely with itself than it is with other species.

CONCLUSIONS

Biological intuition, mathematical models and empirical studies together reinforce one another in leading to the understanding of ecological and evolutionary processes. It is sometimes thought that mathematical models are uncoupled from the others. Their apparent precision and objectivity does not reflect the degree to which they depend on their assumptions. There is too much of a tendency among mathematicians and non-mathematicians alike to adopt what I call the "cafeteria approach" to models. Models and their predictions are chosen because they agree with intuition, not necessarily because the underlying assumptions are especially appropriate. There is nothing wrong with this approach as long as the models are not then used to provide independent support for the same intuition.

ACKNOWLEDGEMENTS

I wish to thank L. Houck, D. Simberloff and M. J. Wade for numerous helpful comments on an earlier draft of this paper.

LITERATURE CITED

BROWN, W.L., JR. and E.O. WILSON. 1956. Character displacement. Systematic Zoology, 5:49-64.

BULMER, M.G. 1974. Density dependent selection and character displacement. American Naturalist, 108:45-58.

CROZIER, R.H. 1974. Niche shape and genetical aspects of character displacement. American Zoologist, 14:1151-1157.

DARWIN, C.R. 1859. On the origin of Species. John Murray, London, 490 pp.

DUNHAM, A.E., G.R. SMITH and J.N. TAYLOR. 1979. Evidence for ecological character displacement in western American catosomid fishes. Evolution, 33:877-896.

FENCHEL, T.M. and F.B. CHRISTIANSEN. 1977. Selection and interspecific competition. *In* F.B. Christiansen and T.M. Fenchel, editors, Measuring Selection in Natural Populations, Lecture Notes in Biomathematics, Vol. 19, Springer-Verlag, New York. Pp. 477-498.

GRANT, P.R. 1972. Convergent and divergent character displacement. Biological Journal of the Linnean Society, 4:39-68.

GRANT, P.R. 1975. The classic case of character displacement. Pp. 237-337. *In* T. Dobzhansky, M.K. Hecht and W.C. Stexere, editors, Evolutionary Biology, Plenum Press, New York.

GRANT, P.R. and T.D. PRICE. 1981. Population variation in continuously varying traits as an ecological genetics problem. American Zoologist, 21:795-811.

HUTCHINSON, G.E. 1959. Homage to Santa Rosalia, or Why are there so many kinds of animals? American Naturalist, 93:145-159.

KIMURA, M. 1965. A stochastic model concerning the maintenance of genetic variability in quantitative characters. Proceedings of the National Academy of Science of the United States, 54:731-736.

LACK, D. 1947. Darwin's Finches. Cambridge University Press, London.

LANDE, R. 1976. The maintenance of genetic variability by mutation in a phenotypic character with linked loci. Genetic Research, Cambridge, 26:221-235.

LAWLOR, L.R. and J. MAYNARD SMITH. 1976. The coevolution and stability of competiting species. American Naturalist, 110:79-99.

MACARTHUR, R. 1968. The theory of the niche. Pp. 159-176. *In* R. C. Lewontin, editor, Population Biology and Evolution. Syracuse University Press, Syracuse, New York.

MACARTHUR, R. and R. LEVINS. 1967. The limiting similarity, convergence and divergence of coexisting species. American Naturalist, 101:377-385.

MAY, R.M. and R.H. MACARTHUR. 1972. Niche overlap as a function of environmental variability. Proceedings of the National Academy of Sciences of the United States, 69:1109-1113.

ROUGHGARDEN, J. 1972. Evolution of niche width. American Naturalist, 106:683-718.

ROUGHGARDEN, J. 1974. Niche width: biogeographic patterns among *Anolis* lizard populations. American Naturalist, 108:429-442.

ROUGHGARDEN, J. 1976. Resource partitioning among competing species--a coevolutionary approach. Theoretical Population Biology, 9:388-424.

ROUGHGARDEN, J. 1979. Theory of Population Genetics and Evolutionary Ecology: An Introduction. MacMillan, New York.

ROUGHGARDEN, J., D. HECKEL, and E.R. FUENTES. 1983. Coevolutionary theory and the biogeography and community structure of *Anolis*. *In* R.B. Huey, E.R. Pianka and T.W. Schoener, editors, Lizard Ecology: Studies of a Model Organism, Harvard University Press, Cambridge, Massachusetts. In press.

SIMBERLOFF, D. and W. BOECKLEN. 1981. Santa Rosalia reconsidered: size ratios and competition. Evolution, 35:1206-1228.

SLATKIN, M. 1979. Frequency- and density dependent selection on a quantitative character. Genetics, 93:755-771.

SLATKIN, M. 1980. Ecological character displacement. Ecology, 61:163-177.

STRONG, D.R., L.A. SZYSKA and D.S. SIMBERLOFF. 1979. Tests of community-wide character displacement against null hypotheses. Evolution, 33:897-913.

TURELLI, M. 1977. Random environments and stochastic calculus. Theoretical Population Biology, 12:140-178.

TURELLI, M. 1978. A reexamination of stability in randomly varying versus deterministic environments with comments on the stochastic theory of limiting similarity. Theoretical Population Biology, 13:244-267.

YABLOKOV, Y.A. 1974. Variability in Mammals. Smithsonian Institution, Washington, D.C.

INDEX

D

F

O

P

S

U

V

Y

Z